Sirozhiddin Zoirov
Mohichehra Shajmardanova

Production of monocalcium from red sand central phosphates

Sirozhiddin Zoirov
Mohichehra Shajmardanova

Production of monocalcium from red sand central phosphates

Monograph

ScienciaScripts

Imprint
Any brand names and product names mentioned in this book are subject to trademark, brand or patent protection and are trademarks or registered trademarks of their respective holders. The use of brand names, product names, common names, trade names, product descriptions etc. even without a particular marking in this work is in no way to be construed to mean that such names may be regarded as unrestricted in respect of trademark and brand protection legislation and could thus be used by anyone.

Cover image: www.ingimage.com

This book is a translation from the original published under ISBN 978-620-8-11742-9.

Publisher:
Sciencia Scripts
is a trademark of
Dodo Books Indian Ocean Ltd. and OmniScriptum S.R.L publishing group

120 High Road, East Finchley, London, N2 9ED, United Kingdom
Str. Armeneasca 28/1, office 1, Chisinau MD-2012, Republic of Moldova, Europe
Printed at: see last page
ISBN: 978-620-8-26543-4

MINISTRY OF HIGHER EDUCATION, SCIENCE AND INNOVATIONS OF THE REPUBLIC OF UZBEKISTAN
TERMEZ INSTITUTE OF ENGINEERING AND TECHNOLOGY

Zoirov Sirojiddin Sakhomiddin ugli,

DEVELOPMENT OF TECHNOLOGY FOR PRODUCTION OF MONOCALCIUM AND MONAKALIY PHOSPHATE ON THE BASIS OF PHOSPHORITES FROM THE CENTRAL KYZYLKUMS

MONOGRAPHY.

Termez-2024

TABLE OF CONTENTS

INTRODUCTION

At present, due to the rapidly growing population of the planet, with the reduction of arable and irrigated lands, provision of the population with food and drinking water becomes more and more acute. Despite huge achievements in agriculture and livestock breeding, this problem remains unsolved by the beginning of the XXI century. One of the most effective ways to solve this problem is to further increase crop yields and productivity of livestock, poultry and fish farming. In this connection, provision of agro-industrial complex with highly effective varieties of grain crops, mineral and organomineral fertilizers, introduction of advanced agrotechnical and agrochemical technologies, application of mineral fodder additives and mixed fodders acquire topical importance, among which monocalcium phosphate and monocalcium phosphate of fodder purity occupy a special place.

In the world, practically in all developed countries, there is a tendency to increase the production and range of chlorine-free potassium and NPK fertilizers, feed calcium phosphates. To provide agriculture with chlorine-free potassium and, on their basis, chlorine-free, fully water-soluble NPK-fertilizers for drip irrigation and foliar feeding of plants, animal husbandry in calcium phosphates of fodder purity, it is necessary to justify a number of solutions: development of effective methods of obtaining chlorine-free potassium fertilizers and monocalcium phosphate of feed purity; study of fluorine content and other impurities during concentration of desulfurized and desulfurized extraction phosphoric acid (EPA) from phosphorites of Central Kyzylkum (CK); establishment of optimal technological parameters of the conversion process of monosodium phosphate and potassium chloride.

The Republic has achieved great success in implementing innovative developments of scientists in the field of production of single and complex phosphate fertilizers and phosphoric acid salts from phosphate rock. The third direction of the development strategy of the Republic of Uzbekistan for 2022-2026 notes important tasks aimed at "... continuing the implementation of industrial policy aimed at ensuring the stability of the national economy, increasing the share of industry in the gross domestic product and increasing the volume of industrial production by 1.4 times"[1] . In this regard, the development of technologies for the production of import-substituting and export-oriented products such as mineral, feed additives monocalcium phosphate on the basis of EPC from phosphate rock and chlorine-free potassium fertilizers - monocalcium phosphate on the basis of flotation potassium chloride produced by JSC "Dehkanabad Potash Plant" become very important.

This dissertation research to a certain extent serves the fulfillment of the tasks stipulated in the Decree of the President of the Republic of Uzbekistan No. UP-60 dated January 28, 2022 "On the Strategy of Development of New Uzbekistan for 2022-2026" and in the Resolutions of the President of the Republic of Uzbekistan No. PP-4265 dated April 3, 2019 "On measures for further reforming and increasing the investment attractiveness of the chemical industry", No. PP-4937 dated December 28, 2020 "On measures to implement the investment program of the Republic of Uzbekistan for 2021-2023", No. PP-4992 dated February 13, 2021 "On measures for further reforming and financial rehabilitation of chemical industry enterprises, development of production of chemical products with high added value" and No. PP-4005 dated November 6, 2018 "On additional measures for further development of the fish farming industry", as well as other regulatory and legal documents adopted in this area.

This research was carried out in accordance with the priority direction of science and technology development in the Republic of VII - "Chemical technologies and nanotechnologies".

The developed technologies of EFC purification from fluorine compounds by precipitation in the form of alkali metal silicofluorides are not acceptable for CK phosphorites due to the low content of acid-soluble silicon compounds.

Technical solutions of monocalcium phosphate production are reduced to the decomposition of phosphorites with initial or evaporated EPC to obtain double superphosphate, the main component of which is monocalcium phosphate, and which contains up to 4% of fluorine and up to 15% of other impurities.

Developments in the field of production of monocalcium phosphate - chlorine-free potassium fertilizer, mainly include neutralization of TPC or EPC with potassium compounds, as which potash or potassium chloride are used. Therefore, there is a need for science-based, new technical solutions for the production of monocalcium and monokalium phosphates from local raw materials, acceptable to the Republic.

CHAPTER I. STATE OF KNOWLEDGE OF MONOCALCIUM AND MONOCALCIUM PHOSPHATES PRODUCTION PROCESSES

§ 1.1. Applications, demand, scale of production of calcium and potassium phosphates.

§ 1.1.1 Calcium phosphate applications, demand, scale of production.

Along with carbon, hydrogen and oxygen, phosphorus and its compounds play an important role in the vital activity of all living organisms and plant life. Phosphorus occupies a special place among chemical elements. It is a part of many minerals, primarily calcium phosphates. In living nature it forms organophosphorus compounds, which serve as carriers of high-energy reactions that ensure the vital activity of living organisms. The role of phosphorus in living nature is unique. There are substitutes for coal, oil or iron, but there is no substitute for phosphorus.

Phosphorus is the most important component of feed rations of livestock, poultry and fish. It is part of nucleic acids, phosphates, phosphoproteins and other compounds, and is a necessary component for building bone tissue. Lack of phosphorus in the diets of farm animals reduces meat and milk productivity, leads to bone diseases and impaired reproduction. To eliminate the deficiency of phosphorus in the body of animals apply mineral feed additives, which are introduced to improve the quality of feed rations. The world assortment of basic mineral feed additives includes more than 10 names. Phosphorus-containing mineral feeds based on calcium, sodium, ammonium phosphates and other chemical components are widely used in livestock, poultry and fish farming.

Calcium phosphates are the most valuable. In feeds where there is a significant amount of calcium and not enough phosphorus, phosphorus-sodium additives are used. Non-protein nitrogen-containing compounds - ammonium phosphates - are used to compensate for protein deficiency in cattle and sheep diets.

Phosphorus and calcium participate in metabolic processes of the organism, cause high efficiency of feed mineral supplements. The quality of feed calcium phosphates is assessed by the content of digestible forms of nutrients in them with a minimum concentration of harmful impurities such as fluorine, lead arsenic, cadmium, mercury. Biological assimilability of phosphorus from feed calcium phosphates - monocalcium phosphate, dicalcium phosphate, tricalcium phosphate is not less than 80%.

Currently, the global consumption of feed calcium phosphates is more than six million tons per year and continues to increase annually. Calcium phosphates are produced in powdered and granulated form, and the share of granulated products is constantly increasing and has already exceeded 70%. This is due to their use in the production of premixes and mixed fodders.

Uzbekistan's demand for feed additives (feed phosphates of ammonium, calcium, sodium, etc.) exceeds more than 100 thousand tons per year and also continues to increase.

The increase in demand for feed phosphates from minerals is due to the widespread absolute rejection of cheaper feed additives derived from bone meal, which is associated with the danger of infection with the "mad cow disease" virus. The danger of this virus has led to an unpredictable increase in demand for feed phosphates from mineral raw materials.

The average annual growth in global feed phosphate consumption is 6%, which is about 2.5 times higher than for phosphate-based fertilizers. The Americas consumes the most feed phosphates with 50%, Asia with 18%, Western Europe with 21% and Central and Eastern Europe with only 9%. The largest average annual growth in feed phosphate consumption is observed in Latin America (Brazil +14%) and Asia (China +10%). Feed phosphate consumption is also expected to grow in Central Asia. It should be noted here that the largest producer of feed phosphates in Asia, Lomon (China), which has a wide range of products and produces up to 600 thousand tons of calcium feed phosphates per year, is fully focused on the domestic market. Brazilian producers of feed phosphates also sell their products almost entirely on the domestic market.

Phosphoric acid and its salts are widely used in the production of mineral fertilizers, in the food industry, medicine, pharmaceuticals, electronics, chemical, textile, glass, aviation, machine-building industries. The main amount of phosphate raw materials is used to produce mineral fertilizers (about 80%), 12% - to produce detergents, 5% - to produce feed phosphates, 3% - to produce special purpose products.

Pure calcium phosphorus salts are used in the food industry, in the baking powder system, in medicine, and in the perfume industry. They are used in the manufacture of bone tissue in dentistry and as a filler in the production of toothpastes.

The most important component of feed rations of livestock, poultry, fish is calcium and phosphorus. In this respect, calcium phosphates are universal mineral feed for farm animals of all kinds in case of phosphorus and calcium deficiency in rations.

Modern industrial methods of livestock production are characterized by the wide use of mineral feed additives, which help to increase productivity, safety of livestock, and reduce feed costs.

Monocalcium phosphate (molecular weight 252) is a single-substituted calcium salt of orthophosphoric acid. Pure monocalcium phosphate in anhydrous form $Ca(H_2 RO)_{42}$ contains 60.65% $P_2 O_5$ and 23.96% CaO, and monohydrate $Ca(H_2 RO)_{42} \cdot H_2 O$ - 56.31% $P_2 O_5$ and 22.25% CaO. According to GOST 23999-80 feed monocalcium phosphate of the 1st and 2nd grades should contain, respectively, not less than 55 and 50% of $P_2 O_5$ soluble in 0.4% hydrochloric acid solution. The product of both grades should contain no more than 18% calcium, 0.2% fluoride, 0.006% arsenic, 0.002% lead, and 4.0% water; the pH of 0.01 M aqueous solution should be at least 3 [25; 272 p]. Mono- and dicalcium phosphates dissolve in water incongruently. Their dissolution in water is accompanied by a reaction:

$$Ca(H_2 RO)_{42} \cdot H_2 O + H_2 O \rightarrow CaNRO_4 + H_3 RO_4$$

In the presence of excess water, monocalcium phosphate dissociates to form dicalcium phosphate and free phosphoric acid.

§ 1.1.2 Applications, demand, scale of potassium phosphate production

In recent years, the problem of providing the world's population with food and drinking water has become increasingly acute. Despite enormous achievements in agriculture and livestock breeding, by the beginning of the 21st century the food supply of the planet's population remains unsatisfactory due to rapid population growth rates, reduction of arable and irrigated land, sharp rise in the cost of energy and fuel sources.

According to the UN, the world's population will exceed 9 billion people by 2050, while currently more than 1 billion people do not receive the minimum amount of food and half of them are chronically malnourished, more than 20% do not have sources of fresh drinking water. This is due to the drastic reduction of land for grain crops. Thus, since the mid-nineties, the world's area has halved from 0.24 to 0.12 hectares per capita. By 2050, according to the UN, it will decrease to 0.08 hectares per capita. In connection with this, there is an acute problem of providing mankind with food. This problem concerns Uzbekistan as well.

Drying up of the Aral Sea and lack of fresh water caused serious damage to agriculture of the Republic. Due to the deficit of water resources, irrigated arable land per capita decreased from 0.22 hectares to 0.13 hectares. The President and the Government of the country pay great attention to restoration of arable and irrigated lands, development of rainfed areas, application of water-saving, advanced agrotechnical and agrochemical technologies, intensification of agricultural production by increasing crop yields through creation of new high-yielding varieties, use of drip irrigation, foliar feeding of plants.

Agrochemical complex is a key component of our country's economy, for intensification of which one of the important factors is, first of all, wide application and effective use of mineral and organomineral fertilizers, microelements, plant growth and development stimulants, biological means of plant protection, modern technologies of crop cultivation.

Introduction of modern, advanced technologies for growing crops in the closed ground, use of drip irrigation impose more stringent requirements to the range and quality of mineral fertilizers: absence of chlorine in the composition, complete solubility in water, content of the main macronutrients - NPK in various ratios. One of the types of complex RK-fertilizers and the main component of water-soluble, chlorine-free NPK-fertilizers is potassium dihydrophosphate, which is suitable for use on any soil and in any ratio with nitrogen and phosphorus fertilizers.

However, potassium dihydrophosphate as a mineral fertilizer is currently practically not produced for widespread consumption, due to the limited raw material base and the lack of developed, acceptable technologies for its production. Therefore, it belongs to expensive and scarce products.

Potassium is one of the three basic elements of plant nutrition and belongs to macronutrients. Its deficiency in the soil leads to a significant reduction in yields and various diseases.

For many years, 75 kg of potassium chloride in terms of K_2O were applied to cotton crops, the main crop of the Republic, according to the recommendations of agrochemists. This rate was underestimated by 1.5-2 times. As a result, due to the lack of potassium, its removal with the crop, annual washing of saline soils, arable land was depleted of potassium. This led to a sharp decrease in raw cotton yield, fiber quality and efficiency of nitrogen and phosphorus fertilizers. At correct application of NRC-fertilizers the yield on non-saline soils increases by 1.9-3.0 c/ha, on saline soils by 2.5-7.0 c/ha. This confirms the need to increase the application rates of potash fertilizers, which led to an increase in demand for potash fertilizers, especially for chlorine-free fertilizers.

Potassium chloride is the world's main potash fertilizer. It is produced by flotation and halurgical methods in powdered and granular form. The presence of high chlorine content in its composition limits its use on crops where chlorine has a strong negative effect.

Chlorine-free potassium fertilizers are especially valuable in the cultivation of potatoes, beets, sunflowers, grapes and other crops. They have a much more effective effect on the yield and quality of agricultural products if they are used in combination with nitrogen, phosphorus and organic fertilizers.

The demand for chlorine-free potash fertilizers has increased sharply with the development of greenhouse farms, hydroponics, drip irrigation and foliar feeding of plants in the cultivation of vegetables, fruits, grapes and other crops as fully water-soluble, chlorine-free potash fertilizers. Among chlorine-free fertilizers, phosphates and potassium sulfate are in great demand as the main components of complex NPK fertilizers.

Single-substituted potassium phosphate KH_2 PO_4 or potassium dihydrophosphate is a colorless, odorless crystals, molecular weight 136.06 g/mol, density 1380 g/cm^3 , melting point 230-250°C. Potassium dihydrophosphate dissolves well in water without decomposition. At 25°C, the solubility of the pure salt, according to various authors, is 19.92-20.09%.

Monokaliy phosphate is a fully water-soluble, chlorine-free, ballast-free, compound fertilizer consisting of two essential plant nutrients and contains 52.16% P O_{25} and 34.60% K_2 O. It is used under all types of crops and on various types of soils as a phosphorus-potassium mineral fertilizer. Monokaliy phosphate is also used in NPK fertilizers as the main component. Good solubility allows to deliver nutrients directly to the root system in the form of aqueous solutions. This saves not only water resources, but also increases the nutrient utilization rate.

Chlorine-free potassium fertilizers are especially valuable in the cultivation of potatoes, beets, sunflowers, grapes and other crops. They have a much more effective effect on the yield and quality of agricultural products if they are used in combination with nitrogen, phosphorus and organic fertilizers.

The country's demand for chlorine-free potash fertilizers exceeds 20,000 tonnes per year. In addition to the needs of the domestic market, there is a strong demand for chlorine-free potassium phosphates in the foreign market.

Potassium compounds are also widely used in other industries, such as ferrous and non-ferrous metallurgy, textile, glass, pharmaceutical, pulp and paper. In the food industry as an actioxidant and bactericidal agent, in risers for baking, in milk powder and cream - as a stabilizer, together with other additives - in cheese production. They are added to soft drinks for athletes, in dairy products (ice cream, condensed milk), desserts, sauces, soups, syrups, in milk products. They are one of the components of detergents (shampoos, soaps), medicines, valuable phosphorus-potassium fertilizer. Despite this, only 5-6% of the output is used in other industries.

From 2012 to 2015, potash fertilizer production increased from 29.1 million tonnes to 31.5 million tonnes K_2O or 4.2%. By 2019, their production increased, compared to 2014, by 9 million tons K_2O or 21%. Potash fertilizer production is increasing not only due to the increase in the capacity of enterprises, but also due to the commissioning of new production facilities.

The need to increase the production of potassium phosphate salts is determined not only by the increasing demand for mineral fertilizers from traditional consumers, but also by the expansion of their application areas.

§ 1.2. Existing methods of calcium phosphate production

The obtaining of feed calcium phosphates can be divided into the following groups:

- Hydrothermal calcination of natural phosphates and thermal defluoridation of double superphosphate.
- The interaction of fine limestone, chalk or dicalcium phosphate dihydrate with thermal phosphoric acid or purified extraction phosphoric acid.
- Conversion of ammonium phosphate by calcium nitrate.

Technologies of feed phosphates are divided by methods of phosphate raw material decomposition (nitrogen-, sulfur-, hydrochloric-acid), retorting process (retorting, non-retorting), thermal, by used component for precipitation (milk of lime, suspension of chalk or limestone, quicklime).

Double superphosphate is a concentrated phosphate fertilizer, the main component of which is monocalcium phosphate. Double superphosphate is obtained by treating natural phosphate with concentrated phosphoric acid.

The presence of a large amount of fluorine (up to 3% and more) in the composition of double superphosphate does not allow its use as a feed additive. To obtain a fodder product, double superphosphate is subjected to defluorination by heat treatment at a temperature of 150-180°C. This is the cheapest way to obtain feed monocalcium phosphate. However, it is practically impossible to obtain more pure calcium phosphate salts from it.

Phosphorus salts of higher qualification are obtained by neutralization of deeply purified thermal phosphoric acid to the corresponding grades "h", "cda", "hh" by appropriate carbonates or metal hydroxides. The chemistry of monocalcium phosphate, dicalcium phosphate and their mixtures can be represented by the following reaction equations:

$$CaCO_3 + 2H_3 PO_4 \rightarrow Ca(H_2 PO)_{42} \cdot H_2 O + CO_2$$

$$CaCO_3 + H_3 PO_4 \rightarrow CaHPO_4 + CO_2 + H O_2$$

$$CaCO_3 + H_3 PO_4 + H_2 O \rightarrow CaHPO_4 \cdot 2H_2 O + H O_2$$

$$Ca(H_2 PO)_{42} \cdot H_2 O + H_2 O \leftrightarrow CaHPO_4 + H_3 PO_4 + H O_2$$

$$Ca(H_2 PO)_{42} \cdot H_2 O + CaCO_3 \rightarrow 2CaHPO_4 + CO_2 + 2H O_2$$

Another way to obtain feed and purer calcium phosphates involves the interaction of dicalcium phosphate dihydrate with purified phosphoric acid. The process of monocalcium phosphate production is described by the following equation:

$$CaNRO_4 \cdot 2H_2 O + H_3 PO_4 \rightarrow Ca(H_2 PO)_{42} \cdot H_2 O + H_2 O$$

Depending on the rate of phosphoric acid, a mixture of monocalcium phosphate and dicalcium phosphate with different ratios can be obtained.

In recent years, research has been conducted on obtaining feed and purer calcium phosphates from EFC, the essence of which consists in purification of acid from fluorine, sulfates, iron, aluminum and other interfering impurities by introduction of alkali metals, partial neutralization with ammonia in the presence of calcium salts, separation of precipitated compounds and separation of dicalcium phosphate.

Until recently, practically the only tested and technologically realized method of processing phosphate raw materials into feed phosphates was the hydrothermal process of high-temperature calcination of natural phosphates. Desfluorination of phosphate raw materials is carried out at a temperature of 1400-1450 ° C in rotary kilns with small additions of SiO_2 or EFK. The decisive role in this process is played by water vapor, which causes the transition of fluorapatite to hydroxyapatite with the release of fluoride compounds in the gas phase. Hydroxyapatite under the action of high temperature decomposes into tricalcium phosphate and tetracalcium phosphate. This method is associated with high energy costs and produces a low-quality product containing 36-38% P_2 O .5

Other disadvantages of this process are difficult sanitary conditions, complicated apparatus design, and low equipment utilization rate. More concentrated feed calcium phosphates (monocalcium phosphate and dicalcium phosphate) are obtained only from high-quality phosphate raw materials - apatite concentrate, or using expensive and scarce electrothermal phosphoric acid.

Superphosphoric acid (65-70% P_2 O_5) is diluted to 40-55% P_2 O_5 , kept at 20-100°C for hydrolysis and mixed with a non-toxic CaO-containing reagent. Dicalcium phosphate is prepared as a dry powder by interaction of phosphoric acid with lime. When phosphoric acid (50% H_3 RO_4) interacts with lime until a dry powder with fluidity is formed. The reaction is carried out in several successive stages and in some cases replacing part of the lime with CaCO .3

The method of obtaining calcium phosphate feed includes decomposition of phosphate raw material by excess phosphoric acid at elevated temperature, defluoridation of the obtained mass and neutralization by easily degradable calcium-containing reagent [67; C. 79-86]. Decomposition is carried out at 80-120°C, and before desulfurization and neutralization the obtained mass is filtered and the filtrate is returned to the decomposition stage to achieve the total H_3 RO_4 250-400% of the stoichiometry.

To obtain primary and secondary acidic calcium orthophosphates, 10-40% neutralization of the first H ion$^+$ of phosphoric acid (which contains 40-55% P_2 O_5 and ≤0.3% F) with phosphate at 90-130°C is carried out. The resulting slurry is treated with calcium carbonate, oxide or hydroxide at 50-100°C until an aqueous solution is obtained which has a pH of 3-6 and the resulting product is dried at 80-150°C.

According to another method of treatment of crushed chalk or limestone is carried out with concentrated phosphoric acid and in the resulting mixture is introduced additionally aqueous suspension of dicalcium phosphate with a concentration of 350-850 g / l at a temperature of 45-90 ° C.

To increase the productivity of livestock, poultry and fish farming as mineral feeds use calcium phosphates, both independently, added to feed, and as part of biofeeds. Calcium phosphates - monocalcium phosphate, dicalcium phosphate, tricalcium phosphate - contain such important elements as phosphorus and calcium.

The advantages of calcium phosphates, compared to bone meal and plant-based feeds, are determined by chemical, physical and biological properties. Quality criteria are:

- high biological digestibility of the product;
- stable nutrient content;
- extremely low levels of heavy metals and fluoride;
- particle size distribution.

Until the end of the twentieth century in the CIS countries, defluorinated phosphates, bone meal, fodder precipitate, and mono- and dinatrium phosphates were used as mineral fertilizers in large quantities.

Bone meal was practically abandoned due to its high cost and unstable composition, low absorption of phosphorus when feeding pigs and birds and possible serious animal diseases. The production of desulfurized phosphates has been discontinued due to the high cost of energy carriers, dicalcium phosphate at the SE "Prikaspinsky Mining and Metallurgical Combine" in Shevchenko, Kazakhstan due to the lack of apatite raw materials. The only production operating in Russia of fodder precipitate JSC "Meleuzovskie Mineralniye Fertilizers" produces products from evaporated apatite EFC (52% P_2O_5) and limestone using a special, closed technology of purification of the product from undesirable impurities.

Since 2000, NIUIF has been working on improving the technology of fodder phosphates - monocalcium phosphate, monodicalcium phosphate, dicalcium phosphate - with the use of pre-purified from fluorine evaporated EFC. Fluorine removal from acid is carried out in the process of its concentration by fire or vacuum evaporation. Chalk or limestone is used as calcium-containing raw material. Displacement of reagents, acid neutralization and granulation are carried out in turbo-paddle mixers, plow-type mixers or in rotating granulators with internal nozzle.

As a result of a complex of physicochemical and technological studies, a method of obtaining feed calcium phosphate in granulated form has been developed, including deep defluorination of phosphoric acid with the addition of active silicon oxide in the amount of 5 kg/t acid, mixing of chalk with purified EFC and granulation in a high-speed mixer TLG-009. The main technological stages (mixing of reagents, granulation and drying) were studied and the optimal parameters of the technological regime were substantiated.

Another method includes mixing of EFC with calcium-containing component in the presence of retort, granulation and drying of the product. EFC with the content of 62-65% $P_2 O_5$ is used, calcium carbonate or its mixture with lime is used as calcium-containing raw material and the process is carried out at retour 1:(0,3-0,5). Then the stage of mixing is combined with granulation in a high-speed mixer with the addition of water to a moisture content of 9.5-13% and drying is carried out at a temperature of 105-115° C.

There is a known method of obtaining feed calcium phosphates, which includes decomposition of tricalcium phosphate EPC and drying of the product at a temperature of 110-150° C. EPC is pre-treated with tricalcium phosphate or other desalting component, the precipitate is separated and acid is taken in the amount necessary to achieve the ratio $P_2 O_{5ЭФК} : P_2 O_{5КФ}$ = 1:(0,5-2,0), the process is carried out at a moisture content of 30-50%, temperature 60-100° C and pelletized at 60-120° C.

A resource-saving technology for the production of defluorinated calcium feed phosphate by optimizing the process, which increases the efficiency of screening by causing segregation phenomenon on the sieve surface, has been developed.

In recent years, methods of monocalcium phosphate production by liquid-phase circulating methods have become widespread. The essence of the method consists in decomposition of phosphate raw material by 3-5 times excess of concentrated 40-65% by $P_2 O_5$ phosphoric acid at temperatures of 60-90°C, crystallization of monocalcium phosphate at cooling and separation from mother liquor. The advantage of the cyclic method is the possibility of obtaining monocalcium phosphate practically from any types of phosphate raw materials.

In the works using solubility diagrams in systems CaO-P O_{25} -H_2 O and CaO-P O_{25} -HCl-H_2 O graphical calculations of the process of obtaining monocalcium phosphate monohydrate under the conditions of recycling of mother liquor for temperature 40°C from phosphorites of Karatau and Central Kyzylkum have been carried out.

The obtained results indicate the expediency of processing low-grade phosphorites with 4.5-5 times excess of phosphoric acid and the possibility of obtaining monocalcium phosphate of high quality.

To establish the optimal conditions for decomposition of phosphate rock Karatau thermal phosphoric acid with a content of 40% P O_{25} in the conditions of non-thickening slurries under the recirculation scheme graphical calculations were carried out and it is shown that at the rate of acid 450-500% of the stoichiometry, contacting time 50-60 minutes, at temperature 90-95 ° C, followed by filtration and separation of insoluble residue, cooling the filtrate to 40 ° C for 90 minutes can be obtained monocalcium phosphate, in which the decomposition factor is 99.0-99.5%. Regeneration of mother liquor is carried out by sulfuric acid extraction with subsequent mixing with a fresh portion of thermal phosphoric acid (TPA) and return to the stage of decomposition of phosphate raw material.

In order to involve low-quality substandard phosphate raw materials of Chilisai deposit into highly concentrated phosphate fertilizers by their decomposition with a large excess of phosphoric acid, the rates and mechanism of the process have been studied. Kinetic parameters of the decomposition process were determined and it was found that the decomposition of Chilisai phosphates proceeds rather quickly (25-30 min), since the phosphate component of phosphates is provided in the form of the mineral kurskite and that after the fourth cycle it is possible to regenerate phosphoric acid and return it to the decomposition stage.

The process of decomposition of washed, annealed phosphorite concentrate (MOPC) CK decomposed to 40.74% P O_{25} EFC by cyclic method was studied depending on the acid rate from 300 to 400% of the stoichiometry and temperature change from 70 to 100°C.

Kinetic studies of the decomposition process of rank phosphorite meal (RFM) (17.37% P O_{25}), MOFC (25.77% P O_{25}) at temperatures of 70, 80, 90 and 100°C, the rate of evaporated EFC with the content of 41.20% and 44.98% P O_{25} established that there is a significant increase in the decomposition coefficient ($K_{разл}$.) RFM in comparison with MOFC from 0.74 to 1.17 times and is associated with the mineralogical and chemical composition of phosphate rock. The activation energy indices indicate high reactivity of RFM and MOFC in phosphoric acid processing regardless of the rate, concentration of EFC and temperature of the process.

EFC standards for RFM and MOFC have been established, which are 400-500% for EFC containing 41.20% $P O_{25}$ and 400-600% for EFC containing 44.98% $P O_{25}$, decomposition time 60 minutes, temperature not exceeding 100%.

One of the real ways of processing low-grade phosphate with the recycling of mother liquor in the production of monocalcium phosphate is the decomposition process using hydrochloric acid. The essence of these processes is that the mother liquor, after separation of monocalcium phosphate, and containing calcium chloride mixed with phosphoric acid and as a recycling solution returned to the stage of decomposition. At the same time hydrochloric acid is released into the solution.

However, all considered methods of monocalcium phosphate production by the cyclic method contain fluorine compounds unacceptable for feed phosphate.

§ 1.3. Analysis of existing potassium phosphate production methods

The main methods of obtaining potassium dihydrophosphate are interactions of phosphoric acid with calcium-containing compounds, ion exchange and conversion methods.

When phosphoric acid interacts with salts, potassium hydroxide, reactions occur:

$$H_3 PO_4 + K_2 CO_3 = 2KH_2 PO_4 + CO_2 + H O_2$$

$$H_3 PO_4 + KCl = KH_2 PO_4 + HCl$$

$$H_3 PO_4 + KOH = KH_2 PO_4 + H O_2$$

Modern production of phosphate salts is usually based on multi-stage processes involving the production of phosphoric acid, its neutralization to the corresponding hydrophosphates, their separation, drying and dehydration. Potash or potassium hydroxide and phosphoric acid are commonly used. As a result, the fertilizer is expensive and the scale of its production is small. Even the agricultural demand for greenhouse fertilizers is not met.

In scientific and technical literature there is information on various methods of obtaining potassium dihydrophosphate: neutralization of phosphoric acid with potash, potassium hydroxide or interaction with potassium chloride with hydrochloric acid distillation, exchange reaction between sodium dihydrophosphate and potassium sulfate or potassium chloride.

Currently, the production of feed, food and reactive phosphoric acid salts is based on the use of thermal phosphoric acid produced by electrothermal method at temperatures of 1600-1800°C. These industries are energy-intensive, which in light of the continuous growth of energy prices leads to an ever-increasing decrease in their profitability.

To obtain pure potassium dihydrophosphate salts, thermal phosphoric, purified EPC are used or the process is carried out in the medium of organic reagents [35; 248 c, 96; P. 6-15].

The method of obtaining potassium dihydrophosphate by interaction of THF with potassium hydroxide is known, in which potassium hydroxide is pre-cleaned from insoluble impurities, poured into a crystallizer and with stirring introduced THF to pH 5.8-6, the solution is cooled to a temperature of 15-20 ° C and separated potassium dihydrophosphate crystals.

PFC and potassium hydroxide are very expensive components and this method has been used so far to produce reactive and food grade potassium dihydrophosphate. A more economical and promising way to produce potassium dihydrophosphate are methods that utilize purified PFC or processes using organic solvents.

Purification of EFC from associated impurities is a complicated and relatively expensive method of acid production. PFC can be purified by evaporation and stripping with superheated steam, precipitation in the form of poorly soluble compounds, ion-exchange, sorption methods, using organic solvents.

Therefore, the most promising methods for obtaining potassium dihydrophosphate are those based on the use of EPC without preliminary purification.

A method of obtaining potassium dihydrophosphate by interaction of EPC with potassium chloride in water-alcohol medium was developed. In another work the interaction of phosphoric acid with potassium chloride was carried out in butanol medium. In this case, potassium dihydrophosphate precipitates, while hydrogen chloride and impurities of phosphoric acid remain in solution.

A method of obtaining potassium dihydrophosphate is described, including interaction of EPC and potassium chloride in a molar ratio of 2:1 at 70-145°C, grinding of the obtained mass, extraction of excess acid with a boiling organic solvent and subsequent drying at 80-100°C. As organic solvents aminol, acetone, N-butanol are used. Known methods, where, as organic solvents use aliphatic amines. These methods involve separation of hydrogen chloride and recirculation of the organic solvent.

There are methods of obtaining potassium dihydrophosphate using crude EPC. The essence of the method consists in neutralization of EPC with potassium carbonate (potash) at stoichiometric rate to pH 3.5-4.5 at 80-90°C, separation of the resulting mixture at 70-80°C and crystallization of potassium dihydrophosphate from the solution at 15-20°C.

A method has been developed in which EPC is used together with TFC, which includes mixing 52-54% of phosphoric acid with potassium chloride at a mass ratio of 2:1 and heating to 265°C. The resulting mixture is dissolved in water until a saturated solution is obtained, the water insoluble impurities are separated by filtration and potassium dihydrophosphate is crystallized.

A known method of obtaining chlorine-free monokaliy phosphate, including neutralization of EPC with potassium carbonate solution to pH 3.8-4.5 at a temperature of 70-78 ° C, separation of the resulting mixture by filtration, subsequent crystallization and separation of the finished product on cooling. The mother liquor is returned to the potassium carbonate dissolution stage.

The ion exchange method of potassium dihydrophosphate production consists in heterogeneous reaction between solid phase cation and liquid phase cationite in columns filled with cationite. When the cationite is treated with phosphoric acid in the columns, the interaction follows the following scheme

$$H_3PO_4 + RK = RK + KH_2PO_4$$

with formation of potassium dihydrophosphate. To restore the cationite, potassium chloride solution is passed through the column according to the following scheme

$$RH + KCl = RK + HCl.$$

The resulting solution from the first stage is evaporated, separated potassium dihydrophosphate crystals on cooling and dried.

This method of obtaining potassium dihydrophosphate has not found wide application, although it is rather simple technologically because of obtaining dilute solutions of potassium dihydrophosphate, the use of THF and the need to utilize solutions of hydrochloric acid.

Condensation methods are based on the interaction of purified solutions of phosphoric acid salts and calcium-containing compounds, mainly potassium chloride. The most widely used phosphoric acid salts are ammonium phosphates, the interaction of which with potassium chloride proceeds by the following reactions:

$$NH\ H_{42}\ PO_4 + KCl \rightarrow KH_2\ PO_4 + NH_4\ Cl$$

$$(NH\ )_{42}\ HPO_4 + KCl \rightarrow K_2\ HPO_4 + 2NH_4\ Cl$$

When sodium dihydrophosphate is used, the process is carried out by reaction:

$$NaH_2\ PO_4 + KCl \rightarrow KH_2\ PO_4 + NaCl$$

These reactions are reversible and depend significantly on the solubility of the salts used. The solubility of potassium dihydrophosphate in water is much lower than the solubility of ammonium and sodium dihydrophosphates, and in their combined presence in saturated solutions at high temperatures, potassium dihydrophosphate will precipitate first on cooling.

Known method of obtaining potassium dihydrophosphate, including interaction of phosphoric acid ammonium solution containing 16-20% $P\ O_{25}$, obtained by leaching ammophos from apatite concentrate, with potassium chloride at a molar ratio of K^+ :NH_4^+ at 95°C equal to (0.75-1):1) until complete dissolution followed by filtration, cooling to 20 ° C and separation of crystals.

The method of obtaining complex fertilizer by interaction of potassium chloride and ammonium phosphate solutions at a molar ratio of (1.0-1.3):1 and temperature of 35-85 ° C with subsequent crystallization of single-substituted potassium and ammonium phosphates at temperatures from 0 to -5 ° C is known.

The method of obtaining complex fertilizer by interaction of phosphoric acid ammonium solution with concentration 16-20% $P_2\ O_5$, obtained by leaching of ammophos with water at temperature 20-60°C and potassium chloride at the ratio K^+ :$NH4^+$ = (0,75-1):1 is known. Crystallization is carried out at a temperature of 10-20 ° C, the resulting precipitate is washed with a solution of phosphoric acid ammonium with a concentration of 16-20% $P_2\ O\ ._5$

On the basis of analysis of the system Na^+ , K^+ //Cl^- , $H_2\ PO_4^-$ - H_2 O at temperatures of 25 and 100°C different variants of potassium dihydrophosphate production by conversion method using sodium dihydrophosphate and potassium-containing salts - sylvinite, potassium chloride were calculated.

It is established that the interaction of ammonium phosphate and potassium chloride in aqueous solutions proceeds with the formation of mixed crystals of potassium ammonium phosphate. Potassium ammonium phosphate was obtained on the basis of evaporated apatite EPC and ammophos, by leaching with water. The precipitation of potassium ammonium phosphate has high moisture content, which varies from 11.5 to 25.9% and is a saturated mother liquor. Since the saturated solution contains chlorine and undesirable impurities, they are transferred into the product during the drying process, contaminating it.

Conclusions from Chapter 1.

From the critical analysis of literary sources it is clear that there is a large number of methods of purification of products to obtain particularly pure inorganic substances. Often, the application of one or another method is limited to its selective purification from certain impurities. The most effective among the described methods, allowing deep and complete purification of various substances, both from anions and cations is the method of recrystallization. In the literature there is a description of the method of deep purification of ammonium phosphate by sequential recrystallization to obtain monoammonium phosphate of reactive grades "h", "cda", "hh". The yield of "hh" grade product is not more than 32% of the initial mass of ammophos.

Analysis of available materials indicates the scientific, practical and economic significance of research aimed at developing technology and obtaining pure calcium phosphate salts from EPC. This is primarily due to the total reduction of PFC production, its high cost and complexity of technology.

Generalization of literature data related to the issues of obtaining feed and purer calcium orthophosphates from phosphate raw materials of various deposits indicates the need to purify the solutions from the associated impurities of phosphate rock and especially from fluorine.

Production of calcium salts is based on acid decomposition of phosphate raw materials and subsequent processing of nitric acid solutions and EFC. The most widespread are methods of ESP purification by organic solvents and precipitation methods of purification are gaining momentum. Purification of EPC by precipitation of existing impurities is safer, easy to implement on the existing equipment.

Analysis of literature data also shows that studies on obtaining calcium phosphates are carried out with the use of EPC and nitrophosphate solutions based on apatite as initial solutions. There are no data on obtaining calcium phosphates from other types of phosphate raw materials, in particular, various types of phosphate rock. The use of organic solvents for purification of initial solutions is undesirable, as there is a problem of purification from organic substances and the technological process becomes more complicated. The phosphorites of CC differ significantly in composition and structure from the known deposits in the world. Technologies have been developed for their processing into various types of phosphate-based fertilizers and EPCs. Acceptable methods of purification and production of pure salts on its basis have not been developed for EPC from CK phosphate rock. The available data are scattered and do not allow creating a technology for processing CK phosphate rock into feed ammonium, calcium phosphates and salts of higher qualification.

Mineral feed additives play a huge role in the development of livestock breeding, poultry farming and fish farming. The world assortment of basic mineral feed additives includes more than 10 names. Calcium, ammonium and sodium phosphates are predominantly used. The most valuable are calcium phosphates. In feeds where there is a significant amount of calcium and not enough phosphorus, phosphorus-sodium additives are used. Non-protein nitrogen-containing compounds - ammonium phosphates - are used to compensate for protein deficiency in cattle and sheep diets.

With the rapid growth of the fish farming industry, the demand for mineral feed additives alternative to ammophos has increased, among which defluorinated monocalcium phosphate is in great demand.

Neither potassium dihydrophosphate nor defluorinated monocalcium phosphate are produced in the Republic. In connection with the above-mentioned, the organization of production of fertilizing chlorine-free potassium dihydrophosphate and monocalcium phosphate of feed purity from phosphorites of CC is a problem that needs to be solved.

CHAPTER II. CHEMICAL AND PHYSICOCHEMICAL METHODS OF RESEARCH

§ 2.1. Characterization of the materials used and methodology of experiments

For experimental work, we used EFC pre-desulfurized with metasilicate and sodium carbonate and desulfurized with washed, burnt phosphate concentrate (MOFC), sodium metasilicate, sodium and calcium carbonates (limestone), calcium oxide, potassium chloride, potassium phosphate, gaseous and aqueous ammonia.

The chemical composition of the raw materials used is given in Table 2.1.

Table 2.1.

Chemical composition of raw materials used in the work

Reagent	Chemical composition, wt. %							
	P_2O_5	SO_3	CaO	MgO	Al_2O_3	Fe_2O_3	F	CO_2
EFC (out.)	18,31	2,32	0,31	1,12	1,36	0,93	1,25	-
Purification. EFC (ob. 1)	17,02	0,23	1,58	0,49	0,38	0,25	0,30	-
Purification. EFC (ob. 2)	16,98	0,20	2,09	0,80	0,38	0,25	0,32	-
MOFC	27,50	3,31	54,46	0,99	1,87	0,70	2,70	-
Limestone	-	-	54,88	0,47	0,21	0,10	-	43,76

EFC was used according to TSh 6.6-21:2018 produced by "Ammofos-Maxam" JSC, and MOFC produced by Kyzylkum phosphorite plant according to O'zDSt 2825:2018. Purified EFC from fluoride and sulfate anions by known methods. Desulfurization and desulfurization of initial EFC was carried out at the rate of MOFC 125% (purified EFC ob.1) and 150% (purified EFC ob.2) of CaO for precipitation of sulfate in the form of gypsum and fluorine in the form of CaF_2 . Metasilicate and sodium carbonate were used for precipitation of fluorine in the form of sodium hexafluorosilicate.

As sodium carbonate we used soda ash according to GOST 5100-85, produced by UP "Kungrad soda plant", limestone from Jizzak deposit, gaseous ammonia according to GOST 6221-90 and aqueous solutions of ammonia according to GOST 9-92 produced by JSC "Maxam-Chirchik".

As initial components were used purified solutions of monosodium phosphate, obtained on the basis of defluorinated EFC from phosphorites of CK by neutralization with soda ash and flotation potassium chloride produced by UP "Dehkanabad potash plant" composition (wt. %): KCl - 95,3; NaCl - 2,97; n.o. - 1,1; H_2 O - 0,43, obtained from sylvinites of Tyubegatan deposit by flotation method and sodium dihydrophosphate solution having composition (wt. %): Na_2 O - 6.13; P O_{25} - 13.92; SO_3 - 0.74; CaO - 0.22; MgO - 0.79; Al O_{23} - 0.026; Fe O_{23} - 0.012; F - 0.026.

The experiments were carried out on a laboratory model installation consisting of a glass reactor equipped with an electromechanical stirrer and installed in a thermostated glass container.

§ 2.2. Methods of performing chemical analyses, determining physical properties and conducting physicochemical studies

Feedstock, intermediates and final products were analyzed for the content of the following components: phosphates, sulfates, fluorides, potassium, sodium, magnesium, calcium, aluminum, iron.

Phosphates were determined by differential photometric method based on the formation of yellow-colored phosphorus-nadium-molybdenum complex and photometric measurement of the optical density of this complex at a wavelength of λ = 430-450 nm relative to the reference solution containing a known amount of P_2 O_5 . Phosphate extraction was carried out with nitric acid solution.

Potassium was determined by the method of formation of insoluble in ethyl alcohol precipitate of potassium perchlorate and determination of its mass. Sodium was determined by flame photometric method, the mass fraction of chlorine was determined by mercurimetric method by titration of chlorides with mercuric nitrate solution in the presence of indicator diphynylcarbozone. Calcium and magnesium were determined by the complexometric method, based on the color change of the indicator (fluorexone when determining calcium and acid chromium dark blue when determining magnesium) in the interaction of calcium and magnesium ions with trilon B.

Sulfates were determined by weight method based on precipitation of sulfates with barium chloride in acidic medium and subsequent weighing of the precipitate. The content of iron and aluminum was determined by the complexometric method, based on the titration of iron with trilon B in the presence of sulfosalicylic acid as an indicator and reverse titration of excess trilon B with zinc sulfate solution and determination of aluminum in the presence of xylenol orange as an indicator. Determination of fluorides was carried out by ionometric method based on measurement of fluoride concentration in solution using fluoride selective electrode without preliminary fluoride extraction.

Water in solid samples was determined by drying in a desiccator to constant weight at 100-105°C.

To characterize the intermediate and final products, we studied their physical and chemical properties: density, viscosity, pH. The density of solutions and slurries was determined using pycnometer PZh-2. Kinematic viscosity of solutions and slurries was measured by glass capillary viscometers VPZh-1 and VPZh-2, pH of solutions and suspensions was determined by electromechanical method.

Identification of composition and properties of initial and intermediate substances, intermediates and products was carried out in addition to chemical analysis and determination of physical properties, by X-ray phase, IR spectroscopic methods. X-ray radiograms of samples were taken on the XRD-6100 apparatus (Shimadzu, Japan) with computer control. CuK_{α} -radiation (β-filter, Ni, 1.54178 tube current and voltage mode 30 mA, 30 kV) and constant detector rotation speed of 4 deg/min with a step of 0.02 deg were used. (ω/2θ-coupling), and the scan angle was varied from 4 to 80 .°

The surface morphology and microstructure of the samples were analyzed using a scanning electron microscope SEM - EVO MA 10 (Carl Zeiss, Germany) with an X-ray spectrometer Aztec Energy Advanced X-Act -Oxford Instruments. This instrument is designed for microscopic analysis of structure and defects, including determination of local elemental composition using energy dispersive spectroscopy. The scanning electron microscope experiments were performed as follows. To carry out the sample preparation process, a metal alloy holder was placed on the microscope slide, over which an aluminum foil with a double-sided adhesive surface was glued. The sample under study was applied to this foil. Then the slide was placed in the working chamber of the microscope, from which the air was pumped out to create a vacuum. For the measurement, an accelerating voltage of 12 kV was applied to the filament, and the working distance was 8.5 mm. Images were obtained at scales from 50 μm.

CHAPTER III. STUDY OF THE PROCESS OF MONOCALCIUM PHOSPHATE PRODUCTION ON THE BASIS OF EXTRACTION PHOSPHORIC ACID FROM PHOSPHORITES OF THE CENTRAL KYZYLKUMS

§ 3.1. Investigation of the evaporation process of defluorinated and desulfurized extraction phosphoric acid

Studies on the production of desulfurized monocalcium phosphate were conducted in a glass reactor equipped with a mechanical stirrer and installed in a thermostat. The main raw material for feed phosphate production at Ammofos-Maxam SA is EPC. EPC was preliminarily purified from sulfates and fluorine (ob. 1 and 2) using MOFC and sodium salts - carbonate and metasilicate. Further EFC was subjected to concentration in a vacuum evaporation unit.

The presence in EFC of increased content of calcium, magnesium, aluminum, iron can adversely affect the defluorination of EFC, binding fluorine into complex or poorly soluble compounds, over which the elasticity of vapor silicicofluoric acid is significantly reduced.

The compositions of concentrated EFC containing 17-60% P_2O_5 are presented in Table 3.1.

The table shows that as the content of P_2O_5 in the evaporated acids increases, the content of other components also increases proportionally (Fig. 1).

Thus, calcium oxide content increases from 1.58% to 5.54% at 60% P_2O_5, magnesium from 0.49% to 1.15%, iron oxide from 0.25% to 0.85%, aluminum oxide from 0.38% to 1.24%, sulfate ions from 0.23% to 0.76%. Fluoride content decreases from 0.30% to 0.14% depending on the concentration of EFC.

Table 3.1.

Chemical composition of packed extraction phosphoric acids from phosphorites of the Central Kyzylkum region

№	Chemical composition, wt. %						
	P_2O_5	CaO	MgO	FeO_{23}	AlO_{23}	SO_4^{2-}	F
Purified EFC (ob. 1)							
1	17,02	1,58	0,49	0,25	0,38	0,23	0,30
2	25,0	2,33	0,72	0,37	0,56	0,34	0,24
3	35,04	3,25	1,01	0,52	0,78	0,48	0,22
4	40,10	3,72	1,44	0,60	0,91	0,55	0,19

5	45,02	4,19	1,62	0,67	1,01	0,61	0,18
6	50,09	4,64	1,80	0,73	1,08	0,66	0,17
7	55,01	5,09	1,98	0,78	1,14	0,70	0,15
8	60,05	5,54	1,15	0,85	1,24	0,76	0,14
Purified EFC (ob. 2)							
9	16,98	2,09	0,80	0,25	0,38	0,23	0,32
10	25,03	2,93	1,17	0,37	0,56	0,34	0,23
11	35,08	4,31	1,64	0,52	0,78	0,48	0,22
12	40,05	4,93	1,88	0,60	0,90	0,54	0,21
13	45,01	5,54	2,11	0,67	1,01	0,61	0,20
14	50,04	6,16	2,36	0,75	1,12	0,68	0,18
15	55,02	6,78	2,59	0,82	1,23	0,75	0,18
16	60,01	7,40	2,83	0,90	1,34	0,82	0,17

Calcium oxide content increases from 2.09% to 7.40% at 60% $P_2 O_5$, magnesium from 0.80% to 2.83%, iron oxide from 0.25% to 0.90%, aluminum oxide from 0.38% to 1.34%, sulfate ions from 0.23% to 0.82%. Fluoride content decreases from 0.32% to 0.17% depending on the concentration of EFC (ob. 2).

With increasing $P_2 O_5$ content in the evaporated EFC the densities and viscosities increase, and with increasing temperature they decrease (Table 3.2). While the initial 17.02% $P_2 O_5$ defluorinated and desulfurized EFC (Eq. 1) has a density of 1.138 g/cm^3 at 20° C, the acid containing 50% $P_2 O_5$ has a density of 1.622 g/cm^3 . Increasing the temperature from 20° C to 100° C causes the density of the acid containing 50% $P_2 O_5$ to decrease from 1.622 g/cm^3 to 1.583 g/cm .^{3}

Changes in viscosity of the evaporated acids are similar to changes in density. The viscosity of the acid containing 17.02% $P_2 O_5$ is 2.155 mPa·s and that of 50% $P_2 O_5$ 41.460 mPa·s at 20° C and decreases to 0.928 mPa·s and 15.734 mPa·s, respectively, at 100° C.

Table 3.2.

Effect of concentration and temperature on density and viscosity of evaporated phosphoric acids

№	Conz-I EFC	Density, g/cm^3					Viscosity, mPa·s				
		20°C	40°C	60°C	80°C	100°C	20°C	40°C	60°C	80°C	100°C
Cleansing EFC (ob. 1)											
1	17,02	1,138	1,128	1,121	1,118	1,110	2,155	1,426	1,087	1,021	0,928
2	25,0	1,269	1,258	1,251	1,247	1,239	3,821	2,527	1,757	1,405	1,194
3	35,04	1,379	1,366	1,358	1,355	1,347	9,101	6,021	4,187	3,346	2,811

4	40,10	1,468	1,455	1,446	1,442	1,434	17,882	11,832	8,227	6,575	5,457
5	45,02	1,551	1,536	1,527	1,523	1,514	29,772	19,698	13,697	10,946	8,866
6	50,09	1,622	1,607	1,597	1,593	1,583	41,640	28,953	23,138	18,510	15,734
Cleansing EFC (ob. 2)											
7	16,98	1,184	1,173	1,166	1,163	1,156	2,252	1,490	1,136	1,037	0,931
8	25,03	1,320	1,308	1,300	1,297	1,291	3,992	2,641	1,836	1,468	1,248
9	35,08	1,434	1,421	1,412	1,409	1,403	9,510	6,292	4,375	3,497	2,937
10	40,05	1,527	1,513	1,504	1,500	1,493	18,687	12,364	8,597	6,871	5,772
11	45,01	1,613	1,598	1,588	1,584	1,577	31,112	20,585	14,313	11,439	9,380
12	50,04	1,687	1,671	1,661	1,657	1,650	43,514	30,256	24,181	19,345	16,442

While the initial 16.98% P_2O_5 desulfurized and desulfurized EFC (Circulation 2) has a density of 1.184 g/cm^3 at 20° C, the acid containing 50% P_2O_5 has a density of 1.687 g/cm^3 . Increasing the temperature from 20° C to 100° C causes the density of the acid containing 50% P_2O_5 to decrease from 1.687 g/cm^3 to 1.650 g/cm . ^{3}

The viscosity of the acid containing 16.98% P_2O_5 is 2.252 mPa·s and that containing 50% P_2O_5 43.514 mPa·s at 20° C and decreases to 0.931 mPa·s and 16.442 mPa·s, respectively, at 100° C.

§ 3.2. Investigation of the influence of technological parameters on the process of monocalcium phosphate production on the basis of desulfurized and desulfurized extracted phosphoric acid and calcium carbonate

To obtain feed monocalcium phosphate from limestone and desulfurized, desulfurized EFC, the effect of temperature and process duration on the degree of decomposition of limestone at an acid rate of 100% and concentration of 17-50% was investigated (Table 3.3).

Increasing the temperature of the decomposition process from 20 to 80°C significantly increases the degree of decomposition of limestone for all values of process duration. With increasing the duration of the decomposition process from 10 to 100 minutes, the degree of decomposition from 18.17-19.46% at 20°C increases to 77.67-82.64%, and at 80°C these values are 54.60-58.53% and 93.54-98.04%, respectively. Increasing the concentration of EFC from 17% P_2O_5 to 50% P_2O_5 leads to a decrease in the degree of decomposition from 82.64-

98.04% to 77.67-92.14% for temperature 20-80°C and process duration 100 min.

From the technological point of view, the most acceptable concentration of EFC is 30-35% P_2O_5 , at which the degree of decomposition of limestone is more than 95%, with a process duration of 100 minutes. At these parameters, the pulp is mobile and with a sufficient degree of decomposition.

Table 3.4 summarizes the effect of limestone particle size as a function of EFC concentration and process duration at an acid rate of 100%. Reducing the particle diameter from 5.0 to 0.1 mm of limestone leads to an increase in the degree of decomposition at 60°C from 49.33% to 69.11% for 30% P_2O_5 EFC and from 47.36% to 65.20% for 45% P_2 O

.5

Table 3.3.

Effect of temperature, concentration of extraction phosphoric acid and process duration on the degree of limestone decomposition

№	t, °C	Degree of decomposition, %						
		10 min.	20 min	30 min	40 min.	60 min.	80 min.	100 min.
EFC concentration 17% P_2O_5								
1	20	19,46	43,72	63,31	74,86	80,26	81,05	82,64
2	40	33,53	59,57	76,35	84,88	87,12	87,94	89,45
3	60	48,17	70,24	87,02	90,28	92,49	93,79	94,59
4	80	58,53	78,81	94,95	96,87	97,25	97,74	98,04
EFC concentration 20% P_2O_5								
5	20	19,27	43,29	62,68	74,12	79,47	80,24	81,82
6	40	33,20	58,98	75,59	84,04	86,29	87,07	88,56
7	60	47,69	69,55	86,15	89,39	91,57	92,86	93,65
8	80	57,95	78,03	94,01	95,91	96,29	96,77	97,07
EFC concentration 25% P_2O_5								
9	20	19,08	42,86	62,05	73,38	78,68	79,44	81,01
10	40	32,87	58,39	74,83	83,20	85,43	86,20	87,68
11	60	47,21	68,85	85,29	88,49	90,65	91,93	92,71
12	80	57,37	77,25	93,07	94,95	95,33	95,80	96,10
EFC concentration 30% P_2O_5								
13	20	18,99	42,65	61,74	73,01	78,29	79,04	80,60
14	40	32,71	58,10	74,46	82,78	85,01	85,77	87,24

15	60	46,97	68,51	84,86	88,05	90,20	91,47	92,25
16	80	57,08	76,86	92,60	94,47	94,85	95,32	95,62
EFC concentration 35% P_2O_5								
17	20	18,88	42,39	61,37	72,57	77,82	78,56	80,12
18	40	32,51	57,75	74,01	82,28	84,50	85,26	86,81
19	60	46,69	68,10	84,35	87,57	89,66	90,92	91,70
20	80	56,74	78,39	92,04	93,90	94,28	94,75	95,05
EFC concentration 40% P_2O_5								
21	20	18,73	42,05	60,88	71,99	77,20	77,93	79,48
22	40	32,25	57,29	73,42	81,62	83,82	84,58	86,12
23	60	46,32	67,56	83,68	86,87	88,94	90,19	90,97
24	80	56,27	77,76	91,30	93,15	93,53	93,99	94,29
EFC concentration 45% P_2O_5								
25	20	18,45	41,42	60,39	71,41	76,58	77,30	78,85
26	40	31,77	56,43	72,83	80,97	83,15	83,90	85,43
27	60	45,63	66,55	83,01	86,17	88,29	89,47	90,24
28	80	55,43	76,59	90,57	92,40	92,78	93,24	93,54
EFC concentration 50% P_2O_5								
29	20	18,17	40,80	59,49	70,34	75,44	76,15	77,67
30	40	31,29	55,87	71,74	79,76	81,91	82,65	84,15
31	60	44,95	65,54	81,77	84,88	86,97	88,13	88,89
32	80	54,60	75,45	89,22	91,02	91,39	91,85	92,14

The best results of limestone decomposition degree are observed at particle diameter less than 1.0 mm and are 94.38-97.05% at ESP concentration of 30% P_2O_5 and process duration of 100 minutes and 93.05-94.12% for acid concentration of 45% P_2O_5.

Thus, the optimal technological parameters of limestone decomposition were established: decomposition temperature - 80-100°C, process duration - 30-60 min, particle size - 0.1-1.0 mm. Increasing the duration of the decomposition process, increasing the temperature and reducing the diameter of limestone particles also contribute to increase the degree of decomposition.

Table 3.4.

Effect of limestone particle diameter, concentration of extraction phosphoric acid and process duration on the degree of limestone decomposition at a rate of 100% EFC

№	Particle diameter, mm	Degree of decomposition, %						
		10 min.	20 min	30 min	40 min.	60 min.	80 min.	100 min.
EFC concentration 20% P_2O_5								
1	5,0	49,97	68,51	84,86	89,05	90,59	91,47	92,25
2	3,0	55,01	72,69	88,73	91,86	92,93	93,40	93,94
3	1,0	59,08	76,86	91,6	94,15	94,85	95,32	95,62
4	0,5	62,67	79,79	93,78	95,71	96,22	96,55	96,74
5	0,18	66,05	82,17	95,17	97,08	97,56	97,84	98,03
6	0,1	70,01	87,10	97,07	97,66	98,15	98,33	98,52
EFC concentration 30% P_2O_5								
7	5,0	49,33	67,62	83,76	87,89	89,41	90,28	91,05
8	3,0	54,29	71,75	87,58	90,67	91,72	92,19	92,72
9	1,0	58,31	75,86	90,41	92,93	93,62	94,08	94,38
10	0,5	61,86	78,75	92,56	94,47	94,97	95,29	95,48
11	0,18	65,19	81,10	93,93	95,82	96,29	96,57	96,76
12	0,1	69,11	85,97	96,75	96,11	96,77	97,05	97,05
EFC concentration 45% P_2O_5								
13	5,0	47,36	64,91	80,41	84,38	85,84	86,67	87,41
14	3,0	52,21	67,69	83,41	89,37	90,05	91,13	91,89
15	1,0	55,01	72,61	88,12	90,21	91,95	92,32	93,05
16	0,5	58,91	74,29	90,03	91,71	92,21	92,73	93,32
17	0,18	61,49	77,23	90,32	91,39	91,87	93,33	93,92
18	0,1	65,20	81,11	91,27	92,35	92,79	93,81	94,12

§ 3.2.2 Effect of rate and concentration of phosphoric acid on monocalcium phosphate production process

Desfluorinated, desulfurized and evaporated to 16.98-45% P_2O_5 acids were used for limestone decomposition. The influence of concentration and rate of EFC on the chemical composition of pulp and desulfurized monocalcium phosphate was investigated.

The effect of EFC rate with a concentration of 16.98-45% P_2O_5 on the composition of monocalcium phosphate slurries is summarized in Table 3.5.

Table 3.5.

Effect of rate and concentration of extraction phosphoric acid on the chemical composition of monocalcium phosphate slurry (ob. 2)

№	Acid rate, %	Conz. EFC, % P_2O_5	Chemical composition, wt. %						
			P_2O_5	CaO	MgO	Fe O_{23}	Al O_{23}	SO_4^{2-}	F
1	95	17	15,76	8,48	0,94	0,23	0,35	0,21	0,30
	100		15,88	8,22	0,94	0,24	0,36	0,22	0,30
	105		15,93	7,94	0,93	0,24	0,37	0,22	0,30
	110		15,98	7,65	0,93	0,24	0,38	0,22	0,30
2	95	25	22,45	11,95	1,35	0,34	0,50	0,31	0,20
	100		22,69	11,60	1,35	0,34	0,51	0,31	0,21
	105		22,79	11,74	1,34	0,34	0,51	0,31	0,21
	110		22,88	10,88	1,33	0,34	0,51	0,31	0,21
3	95	35	30,19	16,25	1,82	0,45	0,68	0,41	0,18
	100		30,64	15,85	1,82	0,46	0,69	0,42	0,19
	105		30,85	15,42	1,81	0,46	0,70	0,42	0,19
	110		31,06	14,96	1,81	0,46	0,70	0,42	0,19
4	95	40	34,16	18,39	2,05	0,51	0,77	0,46	0,17

	100		34,67	17,94	2,04	0,52	0,78	0,47	0,18
	105		34,91	17,45	2,03	0,52	0,79	0,47	0,18
	110		35,15	16,93	2,03	0,52	0,79	0,47	0,18

As the acid rate increases from 95% to 110%, the contents of P_2O_5 and CaO increase slightly, the contents of impurity components increase by tenths of a percent, regardless of the concentration of EFC. At an acid concentration of 25% P_2O_5 P_2O_5 content increases from 22.45% at 95% to 22.88% at 110% for monocalcium phosphate formation. After drying, the P_2O_5 content of monocalcium phosphate changes from 51.97% to 54.99% (Table 3.6). At the same time, the content of CaO is 25.63-28.24% and fluorine 0.24-1.02%.

Table 3.6.

Effect of rate and concentration of extraction phosphoric acid on the chemical composition of monocalcium phosphate (ob. 2)

№	Norma acid, %	Conz. EFC, % P_2O_5	Chemical composition of monocalcium phosphate, wt. %						
			P_2O_5	CaO	MgO	Fe O_{23}	Al O_{23}	SO_4^{2-}	F
1	95	17	52,32	28,16	3,14	0,78	1,17	0,71	0,98
	100		53,23	27,55	3,16	0,79	1,19	0,72	1,00
	105		53,88	26,91	3,17	0,80	1,20	0,73	1,01
	110		54,53	26,27	3,18	0,82	1,22	0,74	1,02
2	95	25	52,18	27,77	3,13	0,78	1,17	0,71	0,47
	100		53,01	27,11	3,15	0,79	1,18	0,72	0,47
	105		53,45	26,37	3,16	0,80	1,19	0,72	0,48

	110		53,88	25,63	3,18	0,81	1,20	0,73	0,48
3	95	35	51,97	27,98	3,12	0,77	1,16	0,70	0,31
	100		53,02	27,44	3,15	0,79	1,18	0,72	0,32
	105		53,98	26,95	3,18	0,81	1,20	0,73	0,32
	110		54,94	26,46	3,20	0,82	1,22	0,74	0,33
4	95	40	52,22	28,11	3,14	0,78	1,16	0,71	0,27
	100		53,04	27,45	3,16	0,79	1,18	0,72	0,27
	105		53,95	26,94	3,18	0,80	1,20	0,73	0,28
	110		54,86	26,43	3,20	0,82	1,22	0,74	0,28
5	95	45	52,47	28,24	3,15	0,78	1,17	0,71	0,24
	100		53,01	27,43	3,16	0,79	1,19	0,72	0,25
	105		54,02	26,96	3,18	0,80	1,21	0,73	0,25
	110		54,99	26,49	3,20	0,82	1,23	0,74	0,26

The studies have shown the possibility of obtaining defluorinated fertilizer monocalcium phosphate on the basis of defluorinated and desulfurized EPC from phosphate rock. Monocalcium phosphate obtained at EPC concentration of 25-40% P_2O_5 contains 51,97-54,99% P_2O_5 , 25,63-28,24% CaO. The fluorine content is 0.24-1.02%. The higher the concentration of initial EFC, the lower the fluorine content of monocalcium phosphate. Increasing the concentration of EFC up to 40 % P_2O_5 contributes to the reduction of fluorine content in the obtained monocalcium phosphate. However, the content of fluorine in the final product is more than 0.2 %, which does not meet the requirements for feed calcium phosphate. Obtained monocalcium phosphate, above-mentioned concentrations of EFC is defluorinated monocalcium phosphate fertilizers.

To obtain monocalcium phosphate of feed purity, it is necessary to carry out deeper purification of EPC from phosphate rock of CK from fluorine.

For the decomposition of limestone was used defluorinated, desulfurized and steamed content from 45 to 60% P_2O_5 acid. The effect of phosphoric acid concentration and rate on the chemical composition of slurry and desulfurized monocalcium phosphate was investigated.

The effect of EFC rate with a concentration of 45-60% P_2O_5 on the composition of monocalcium phosphate slurries is shown in Table 3.7.

As the acid rate increases from 95% to 110%, the contents of P_2O_5 and CaO increase slightly; the contents of impurity components increase by tenths of a percent, regardless of EFC concentration. At an acid concentration of 45% P_2O_5 the P_2O_5 content increases from 38.46% at 95% to 39.59% at 110%, at an acid concentration of 50% P_2O_5 the P_2O_5 content increases from 42.31% at 95% to 43.56% at 110%, at an acid concentration of 55% P_2O_5 the content of P_2O_5 increases from 46.54% at a rate of 95% to 47.92% at a rate of 110% and at an acid concentration of 60% P_2O_5 the content of P_2O_5 increases from 50.77% at a rate of 95% to 52.27% at a rate of 110% to form monocalcium phosphate.

Table 3.7

Effect of rate and concentration of extraction phosphoric acid on the chemical composition of monocalcium phosphate before drying (Fig. 1)

№	Acid rate, %	Chemical composition, wt. %							
		P_2O_5	CaO	MgO	FeO_{23}	AlO_{23}	SO_4^{2-}	F	H_2O
EFC concentration - 45% P_2O_5									
1	95	38,46	20,69	1,19	0,85	1,26	0,63	0,131	24,67
2	100	39,09	20,23	1,20	0,86	1,27	0,64	0,136	25,32
3	105	39,35	19,65	1,21	0,87	1,28	0,65	0,142	26,43
4	110	39,59	19,06	1,22	0,88	1,29	0,66	0,149	27,49
EFC concentration - 50% P_2O_5									

5	95	42,31	22,77	1,31	0,94	1,39	0,69	0,140	18,14
6	100	43,01	22,25	1,32	0,95	1,40	0,70	0,146	18,65
7	105	43,29	21,61	1,33	0,96	1,42	0,71	0,153	19,47
8	110	43,56	20,97	1,34	0,97	1,43	0,72	0,160	20,25
EFC concentration - 55% P_2O_5									
9	95	46,54	24,50	1,55	1,03	1,53	0,76	0,132	11,09
10	100	47,31	24,47	1,56	1,05	1,54	0,77	0,138	11,56
11	105	47,62	23,77	1,57	1,06	1,56	0,78	0,145	12,48
12	110	47,92	23,07	1,59	1,07	1,57	0,79	0,152	13,25
EFC concentration - 60% P_2O_5									
13	95	50,77	27,32	1,80	1,13	1,67	0,83	0,123	4,86
14	100	51,61	26,70	1,81	1,14	1,68	0,84	0,129	5,23
15	105	51,95	25,93	1,82	1,15	1,70	0,85	0,135	5,98
16	110	52,27	25,16	1,84	1,16	1,72	0,86	0,141	6,31

After drying, the P_2O_5 content in monocalcium phosphate changes from 52.58% to 55.26% (Table 3.8). At the same time, the CaO content is 26.59-28.30% and fluorine content is 0.127-0.181%.

Increasing the concentration of EFC contributes to the reduction of fluoride content in the finished feed purity product.

Table 3.8.

Effect of acid rate and concentration on the chemical composition of monocalcium phosphate after drying (Fig. 1)

№	Acid rate, %	Chemical composition, wt. %						
		P_2O_5	CaO	MgO	Fe_2O_3	Al_2O_3	SO_4^{2-}	F
EFC concentration - 45% P_2O_5								
1	95	52,58	28,30	3,17	1,17	1,72	0,85	0,181

2	100	53,12	27,51	3,15	1,18	1,73	0,86	0,187
3	105	54,11	27,02	3,18	1,20	1,78	0,89	0,192
4	110	55,08	26,53	3,22	1,23	1,81	0,91	0,196
EFC concentration - 50% P_2O_5								
5	95	52,65	28,33	3,17	1,17	1,72	0,85	0,174
6	100	53,19	27,52	3,15	1,18	1,73	0,86	0,179
7	105	54,17	27,04	3,18	1,20	1,78	0,89	0,186
8	110	55,14	26,55	3,22	1,23	1,81	0,91	0,191
EFC concentration - 55% P_2O_5								
9	95	52,73	28,09	3,17	1,17	1,73	0,86	0,144
10	100	53,26	27,55	3,16	1,18	1,74	0,87	0,151
11	105	54,18	27,05	3,16	1,20	1,77	0,89	0,164
12	110	55,21	26,58	3,21	1,23	1,81	0,91	0,173
EFC concentration - 60% P_2O_5								
13	95	52,81	27,95	3,18	1,18	1,73	0,86	0,127
14	100	53,33	27,58	3,16	1,18	1,74	0,87	0,132
15	105	54,24	27,07	3,18	1,20	1,77	0,89	0,140
16	110	55,26	26,59	3,22	1,22	1,81	0,90	0,148

Thus, the conducted studies have shown the possibility of obtaining granulated, feed monocalcium phosphate on the basis of desulfurized and desulfurized EFC from phosphate rock of CK. Monocalcium phosphate regardless of initial acid concentration contains 52,65-55,26% P_2O_5, 26,55-28,33% CaO. The fluorine content is 0.127-0.191%. The higher the concentration of initial phosphoric acid, the lower the fluorine content of monocalcium phosphate. The obtained samples of feed monocalcium phosphate meet the requirements for feed phosphates GOST - 23999-80.

§ 3.3. Influence of retort multiplicity on chemical composition and properties of monocalcium phosphate

When monocalcium phosphate is obtained using the evaporated acid containing 45-55% P_2O_5, a thick mass is formed which sets after 10-15 minutes. The moisture contents of the products are 24.67-27.49%, 18.14-20.25% and 11.09-13.25%, respectively, for acid concentration of 45, 50 and 55% P_2O_5.

Drying of monocalcium phosphate with high moisture content is not economically justified. Therefore, in order to reduce the moisture content of the product fed for drying, the effect of the ratio of MCP:retort on the change in the chemical composition and moisture content of the product was studied. The obtained results are summarized in Table 3.9.

Table 3.9.

Effect of process retouring and phosphoric acid concentration

№	ICF:retour	Chemical composition, wt. %							
		P_2O_5	CaO	MgO	Fe_2O_3	Al_2O_3	SO_4^{2-}	F	H_2O
EFC concentration - 45% P_2O_5									
1	1:0,3	42,22	21,85	1,29	0,93	1,37	0,69	0,153	21,70
2	1:0,5	43,61	22,57	1,34	0,96	1,42	0,72	0,156	19,45
3	1:0,8	45,11	23,35	1,39	0,99	1,47	0,74	0,160	16,86
4	1:1,0	45,88	23,74	1,41	1,01	1,49	0,75	0,162	14,41
EFC concentration - 50% P_2O_5									
5	1:0,3	45,23	23,40	1,39	1,00	1,47	0,74	0,146	17,28
6	1:0,5	46,22	23,91	1,42	1,02	1,50	0,75	0,151	16,08
7	1:0,8	47,30	24,47	1,45	1,04	1,54	0,77	0,155	14,62
8	1:1,0	47,83	24,75	1,47	1,05	1,56	0,78	0,159	13,15
EFC concentration - 55% P_2O_5									
9	1:0,3	48,54	25,11	1,60	1,08	1,58	0,79	0,141	10,97

10	1:0,5	49,09	25,39	1,62	1,09	1,60	0,80	0,143	10,19
11	1:0,8	49,68	25,69	1,64	1,10	1,62	0,81	0,145	9,25
12	1:1,0	49,98	25,85	1,65	1,11	1,63	0,82	0,145	8,75

Increasing the MCF:retort ratio from 1:0.3 to 1:1 or increasing the retort from 0.3 to 1 $P_2 O_5$ content increases from 42.22% to 45.88%, CaO from 21.85% to 23.74% for an acid concentration of 45%, from 45,23% to 47.83% $P_2 O_5$, from 23.40% to 24.75% CaO for an acid concentration of 50% $P_2 O_5$ and from 48.54% to 49.68% $P_2 O_5$, from 25.11% to 25.85% CaO for an acid concentration of 55% $P_2 O_5$. The contents of magnesium, iron, aluminum, sulfate, and fluorine oxides change less significantly and are MgO 1.29-1.65%, Fe O_{23} 0.93-1.11%, Al O_{23} 1.37-1.63%, SO_4^{-2} 0.69-0.82%, and fluorine 0.141-0.162%, while the moisture content of the products is depending on the retort 14.41-21.70% at an acid concentration of 45% $P_2 O_5$, 13.15-17.28% at an acid concentration of 50% $P_2 O_5$ and 8.75-10.97% at an acid concentration of 55% P_2 O .$_5$

Studies have shown that drying of pulp at 100-110°C allows to obtain monocalcium phosphate with the content of 52,65-55,26% $P_2 O_5$, 26,55-28,33% CaO and 0,127-0,191% fluorine.

§ 3.4. Technological scheme and material balance of monocalcium phosphate production on the basis of extraction phosphoric acid and calcium carbonate

On the basis of the obtained results the technological scheme was developed and the material balance of production of fertilizer and fodder monocalcium phosphate from desulfurized, desulfurized and evaporated EFC obtained from phosphate rock.

Figure 3.3 shows the technological scheme for the production of defluorinated, fertilizer and feed monocalcium phosphate [145].

In order to obtain fertilizer monocalcium phosphate, desulfurized and desulfurized EFC is fed to the evaporator (pos. 1) and evaporated to the content of $P_2 O_5$ 30-40%. Evaporated acid is fed to neutralization with limestone in the reactor (pos. 3), the resulting mass is sent to the drying drum (pos. 7) simultaneously with the retort, further products are fed to the cooling drum (pos. 8), classified (pos. 9), large fractions are crushed in the crusher (pos. 10) and returned to classification.

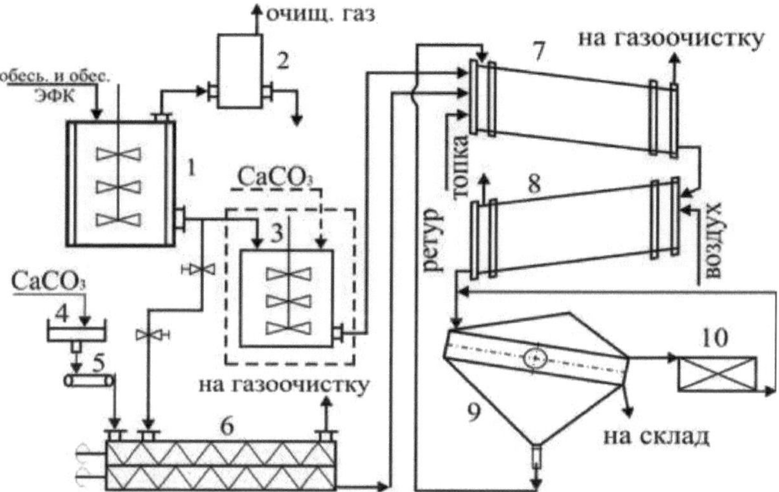

Fig. 3.3. Principal technological scheme of production of defluorinated fertilizer and feed monocalcium phosphate on the basis of extract phosphoric acid obtained from phosphorites of Central Kyzylkum: 1 - evaporator; 2 - absorber; 3 - reactor; 4 - hopper; 5 - flow meter; 6 - screw mixer; 7 - drying drum (BGS); 8 - cooling drum; 9 - classifier; 10 - crusher.

To obtain feed monocalcium phosphate, defluorinated and desulfurized EFC is fed to the evaporator (pos. 1) and evaporated to the content of P_2O_5 45-60%. Evaporated acid is dosed through the flow meter (item 4) together with limestone into the screw mixer (item 5), the resulting mass is fed into the drying drum (item 7) simultaneously with the retort, further products are directed to the cooling drum (item 8), classified (item 9), large fractions are crushed in the crusher (item 10) and returned to classification. Finished products are delivered to the warehouse.

In Figures 3.4 and 3.

5 shows material flows and material balances of monocalcium phosphate production by retort and non-retort methods.

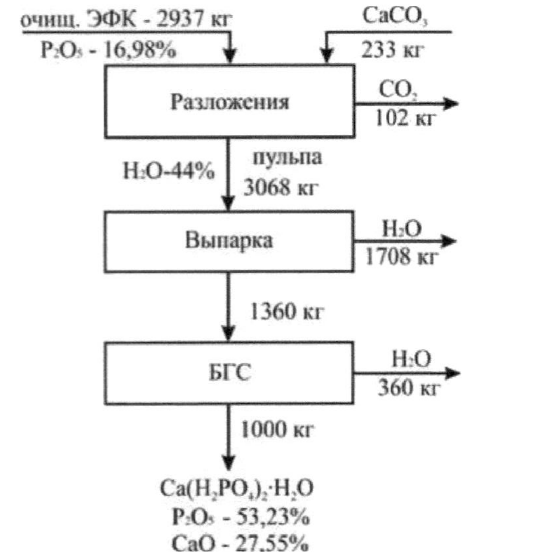

Fig. 3.4. Material balance of granulated monocalcium phosphate production (P_2O_5 16.98%)

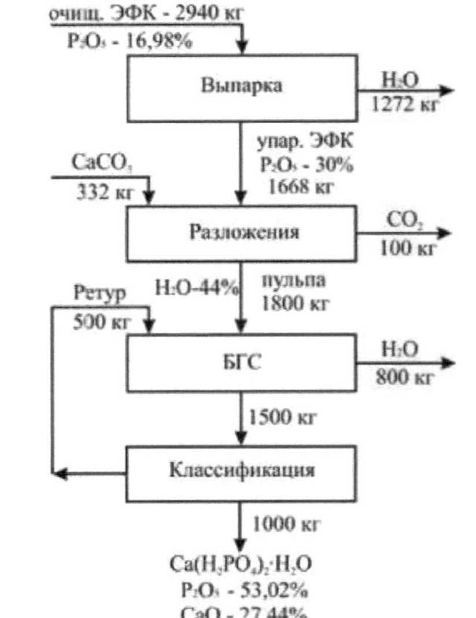

Figure 3.5. Material balance of obtaining granulated monocalcium phosphate in the presence of retort (P_2O_5 30%)

To obtain 1000 kg of monocalcium phosphate it is necessary to evaporate 2940 kg of purified EFC to the content of 30% P_2O_5 and decompose 332 kg of calcium carbonate or 2937 kg of 16,98% P_2O_5 EFC to evaporate and feed into the BGS apparatus for granulation and drying. This produces granulated, defluorinated fertilizer monocalcium phosphate.

To obtain feed monocalcium phosphate it is necessary to use more concentrated EFC. Figure 3.6 shows the scheme of material flows and material balance of feed monocalcium phosphate production from desulfurized and desulfurized EFC at a rate of 100% [145; https://7universum.com/ru/tech/archive/item/12471].

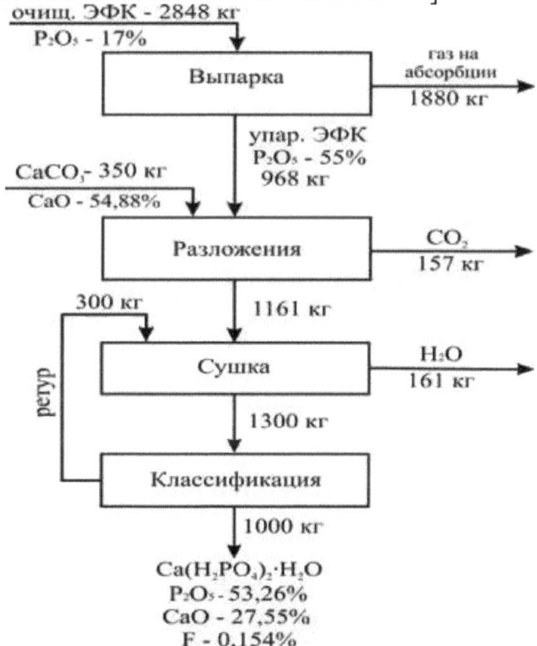

Fig. 3.6. Material balance of feed monocalcium phosphate production on the basis of EFC (P_2O_5 50%)

To obtain feed monocalcium phosphate, which meets GOST 23999-80, first grade, it is necessary to 2848 kg of desulfurized and desulfurized EFC or 968 kg of evaporated EFC decompose 350 kg of limestone, for drying feed 1161 kg of wet monocalcium phosphate and 300 kg of retort (MCP:retort = 1:0.3). This results in 1000 kg of feed monocalcium phosphate after drying. Table 3.10 shows the norms of technological mode of monocalcium phosphate production.

Table 3.10.

Norms of technological mode of monocalcium phosphate production

№	Name of parameters	Values
1. Obtaining ICF using the BGS apparatus		
Desfluoridation and desulfurization of EFCs		
1	Concentration of initial EFC, wt. % P_2O_5	17-21
Uparka desulphation and defluoridation of EFCs		
2	EFK concentration, wt. % P_2O_5	25-40
Decomposition of calcium carbonate by defluorinated and desulfonated EFCs		
3	Limestone content, $CaCO_3$ %	96-98
4	Limestone particle diameter, mm	0,1-1,0
5	EFC rate, wt. %	100-103
6	Decomposition process temperature, °C	80-100
7	Slurry residence time in the reactor, min	30-60
Granulation and drying of monocalcium phosphate		
8	Residual moisture content in pulp, %	25-35
9	Ratio of retour, retour:monocaecium phosphate slurry	(0,4-1,0):1
10	Drying process temperature, °C	100-110
11	Drying time, min.	20-45
2. Obtaining ICFs using the BG and BS apparatus		
Decomposition of calcium carbonate by defluorinated and desulfonated EFCs		
1	EFK concentration, wt. % P_2O_5	45-55
2	EFC rate, wt. %	100-103
3	Decomposition process temperature, °C	80-100
4	Slurry residence time in the reactor, min	30-60
Granulation and drying of monocalcium phosphate		

5	Residual moisture content in pulp, %	6-20
6	Ratio of retour, retour:monocaecium phosphate slurry	(0,2-0,4):1
7	Drying process temperature, °C	100-110
8	Drying time, min.	30-60

Thus, on the basis of the obtained scientific results the norms of technological mode are established, the technological scheme of production is developed and preliminary technical and economic calculations are carried out.

§ 3.5. Physicochemical and commercial properties of the obtained monocalcium phosphate

To conduct research by physicochemical methods of analysis was obtained feed monocalcium phosphate from calcium carbonate and defluorinated, desulfurized, evaporated to 45-55% P_2O_5 EFC at the norm - 95-100%. To verify the purity of the obtained feed monocalcium phosphate, X-ray radiographs and IR spectra were taken (Figures 3.7 and 3.8).

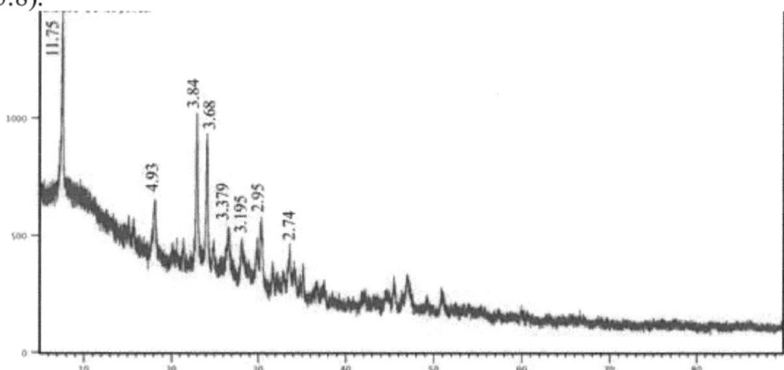

Figure 3.7. Radiograph of feed monocalcium phosphate

In the X-ray radiogram (Fig. 3.7) there are only diffraction maxima characteristic of monocalcium phosphate with interplanar distances of 11.75, 4.93, 3.379, 3.195, 2.95 Å $Ca(H_2RO)_{42}\cdot H_2O$, and, 3.84 and 3.68 Å $Mg(H_2RO)_{22}\cdot 6H_2O$ and $NaH_2PO_4\cdot H_2O$.

On the IR - spectrum (Fig. 3.8) there are frequencies of vibrations characterizing vibrations related to RO_4 440.54-1077.68 cm^{-1} and crystalline water - 1647.11-2897.93 cm $.^{-1}$

Figure 3.8. Infrared spectrum of feed monocalcium phosphate

Figure 3.9 and Table 3.11 show the main components of the obtained monocalcium phosphate based on desulfurized and desulfurized

EPC from CK phosphorites and limestone with in-line method [155; P. 70].

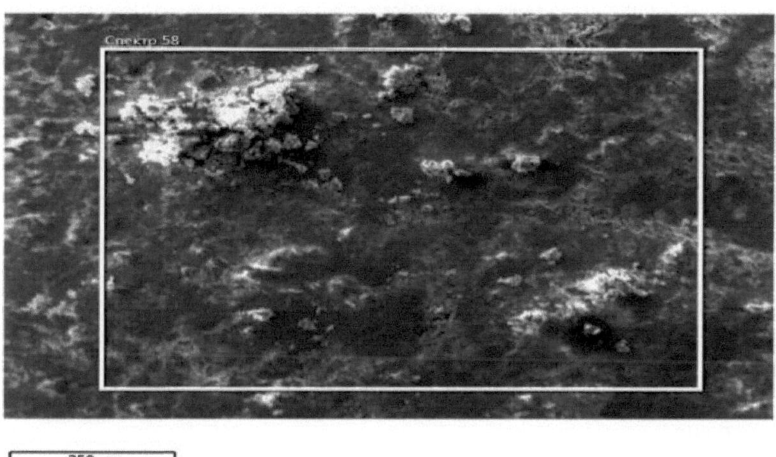

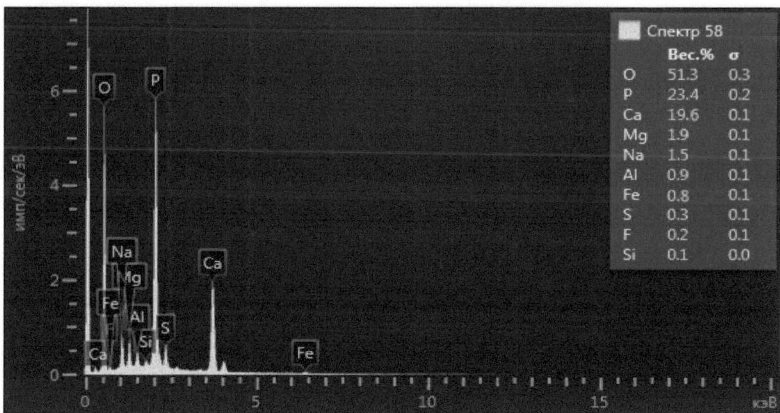

Figure 3.9. Scanning microscopic analysis feed monocalcium phosphate

Table 3.11.

Chemical composition of monocalcium phosphate

Element	Weight. %	Sigma, Wt. %
O	51.28	0.29
F	0.11	0.06
Na	1.51	0.09
Mg	1.89	0.07
Al	0.92	0.05
Si	0.11	0.03
P	23.38	0.22
S	0.31	0.07
Ca	19.68	0.15
Fe	0.81	0.08
Summation:	100.00	

Scanning microscopic analysis of monocalcium phosphate shows the following content of composition elements: O-51.28%, F-0.11%; Na-1.51%; Mg-1.89%; Al-0.92%; Si-0.11%; P-23.38%; Ca-19.68%, S-0.31%, which corresponds to their content in feed monocalcium phosphate.

Thus, the conducted studies have established the possibility of obtaining feed monocalcium phosphate by decomposition of preliminary desulfurized, desulfurized and evaporated calcium carbonate EFC under the following optimal conditions: EFC concentration - 45-55% P_2O_5 , calcium carbonate rate - 95-100%, retort ratio 1.0:0.3-1.0, drying temperature not more than 100-110°C.

§ 3.6 Investigation of the process of crystalline monocalcium phosphate production on the basis of extracted phosphoric acid at high rate

In order to obtain desulfurized monocalcium phosphate in crystalline form, without extraneous impurities, the process of decomposition of limestone by desulfurized and desulfurized EFC from phosphorites CK, pre-propagated to the content of 40-55% P_2O_5 at its rate of 300-500% of stoichiometry for the formation of monocalcium phosphate was investigated.

The process was studied in a laboratory setup consisting of a reactor, mechanical stirrer and thermostat at a temperature of 95-100°C and a process duration of 3 hours. After reaching the set time, the phosphate mass was filtered at the experiment temperature to separate insoluble residue, the filtrate was cooled to 60-70°C and crystalline monocalcium phosphate was separated, washed with water and dried at 100-110°C.

The process of obtaining monocalcium phosphate, by decomposition of limestone with concentrated phosphoric acid, follows the well-known equation:

$$CaCO_3 + 2H_3 RO_4 = Ca(H_2 RO)_{42} + H_2 O + CO_2$$

The mother liquor, after separation of monocalcium phosphate crystals, contains calcium salt dissolved in phosphoric acid, and the insoluble residue also contains undecomposed limestone.

The block diagram of the circulation method of obtaining crystalline, desulfurized monocalcium phosphate is presented in Figure 3.10.

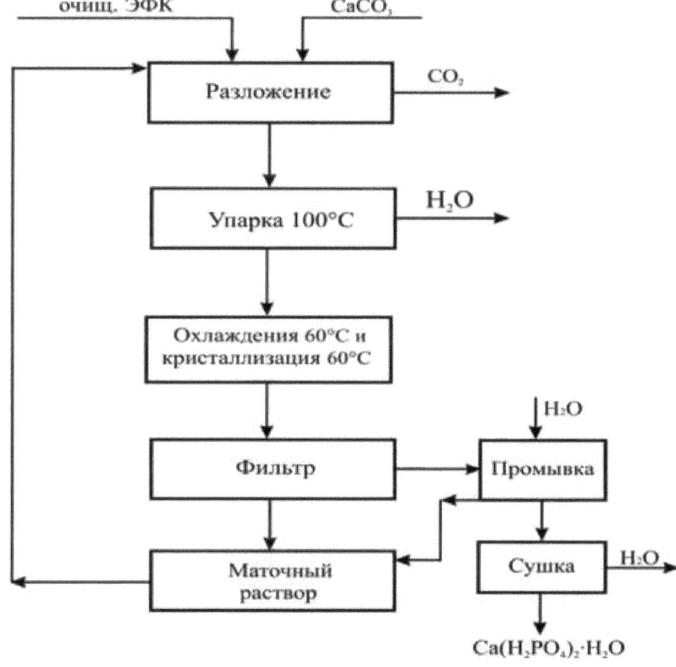

Fig. 3.10. Block diagram of crystalline monocalcium phosphate production on the basis of evaporated EFC CK and limestone at the rate of 400%

According to this scheme, crystalline monocalcium phosphate was obtained at rates of 300-500% of packed extract phosphoric acid containing 45, 50 and 55% P_2O_5 (Table 3.12).

Phosphoric acid mass of decomposition of limestone 45% by P_2O_5 phosphoric acid at rates of 300 and 400% of stoichiometry is practically not filtered. A similar pattern is observed at a rate of 55% phosphoric acid 300%. The best filtration results are observed when 50% P_2O_5 phosphoric acid is used at rates of 300-500% and when 55% P_2O_5 phosphoric acid is used at 500%. In this case, the sludge removal of monocalcium phosphate is 330-450 kg/m^2 ·h and the $P_2O_{5общ.}$ content is 53.6-54.8%, $P_2O_{5св.}$ 11.2 - 14.5%, CaO 16.6 - 17.5%.

Table 3.12.

Influence of the concentration of evaporated extraction phosphoric acid on the quality of monocalcium phosphate (temperature 60°C)

Norm H_3RO_4, % of stx.	H_3RO_4, % P_2O_5	Content, %			Filtration rate		
		P_2O_5		CaO	Duration, sec	Sludge weight, g	Sludge removal, kg/m^2 - h
		Common.	Swob.				
300	45	52,6	16,1	16,7	unfiltered		
400		52,8	12,0	18,9	unfiltered		
300	50	53,6	14,5	16,7	240	17,5	330
400		53,9	13,7	17,3	240	17,5	330
500		54,3	11,3	17,5	250	17,9	380
300	55	53,8	15,9	14,6	unfiltered		
400		55,7	12,0	15,3	280	17,2	180
500		54,8	11,2	16,6	130	17,8	450

Table 3.13 shows the results of the study of the effect of cooling temperature on the crystallization process of monocalcium phosphate, its composition and filtration rate.

The table shows that the cooling temperature significantly affects the process of monocalcium phosphate crystal formation. At temperature of filtrate cooling up to 40°C small crystals are formed, which are poorly filtered or practically not filtered. Good results on filtration of monocalcium phosphate are achieved when cooling the filtrate to a temperature of 60-100°C. Filtration rates under these conditions are for 45% P_2O_5 470-550 kg/m^2 ·h, for 50% P_2O_5 680-820 kg/m^2 ·h, 55% P_2O_5 760-960 kg/m^2 ·h for phosphoric acid concentration. The $P_2O_{5общ}$ content varies from 53.2% to 55.1%, the free form of P_2O_5 is 19.0-4.6%, and the CaO content is 16.2-19.2%.

Table 3.13

Effect of the concentration of evaporated EFC on the quality of monocalcium phosphate

T, °C	Conz. N_3RO_4, % P_2O_5	Content, %			Filtration rate		
		P_2O_5		CaO	Duration, sec	Sludge weight, g	Sludge removal, kg/m^2 - h
		Com mon.	Swob.				
40	45	51,3	21,2	13,7	unfiltered		
60		52,8	12,0	18,9	unfiltered		
80		53,2	18,3	16,2	180	19,0	470
100		53,8	19,0	17,2	120	18,6	550
60	50	53,9	13,7	17,3	240	17,5	230
80		54,2	12,0	18,5	45	28,0	680
100		54,7	9,4	18,9	30	27,6	820
60	55	53,8	15,9	14,6	unfiltered		
80		55,1	15,4	16,9	30	30,0	760
100		54,8	4,6	19,2	20	27,5	960

High content of free P_2O_5 does not allow to use monocalcium phosphate for mixed fodders and as a calcium-phosphate additive to animal, bird and fish feed. To reduce the acidity of monocalcium phosphate it was neutralized with limestone (Table 3.14).

When neutralizing crystalline monocalcium phosphate, obtained using 50% of P_2O_5 EFC at a rate of 400% limestone in the amount of 10 and 15% of the mass of monocalcium phosphate content of free P_2O_5 decreases from 13.7% to 5.2% when crystallizing monocalcium phosphate at 60 ° C and from 12.0% to 4.7% when crystallizing monocalcium phosphate at 80 ° C. The pH of 10% solution of neutralized, crystallized monocalcium phosphate increases from 2.6 to 3.3 and from 2.5 to 3.5.

Table 3.14

Effect of calcium carbonate on acidity and marketability of monocalcium phosphate

$CaCO_3$, % of mass	Conz. N_3RO_4, % P_2O_5	Norm H_3RO_4, % of stack.	°C	pH	Content, %		
					P_2O_5		CaO
					Common.	Swob.	
0	50	400	60	2,6	53,3	13,7	17,3
10				3,1	55,6	10,3	18,4
15				3,3	54,2	5,2	20,3
0	50	400	80	2,5	54,2	12,0	18,5
10				3,2	56,0	9,8	18,5
15				3,5	54,9	4,7	21,0
0	55	300	60	2,3	54,0	16,1	13,8
10				3,0	52,2	6,6	15,4
15				3,25	51,3	3,1	16,9
0	55	400	60	2,2	56,9	20,0	12,5
10				3,0	55,6	8,7	16,3
15				3,1	53,5	5,4	17,5
20				3,3	52,1	2,5	18,7
0	55	400	80	2,2	56,8	19,8	12,1
10				2,8	54,1	8,6	16,5
15				3,2	53,5	5,3	18,3
20				3,5	52,3	2,6	19,8

When crystalline monocalcium phosphate obtained using 55% by P_2O_5 phosphoric acid was neutralized, the content of free P_2O_5 decreased from 16.1% to 3.1% at an acid rate of 300% and a crystallization temperature of 60°C.

At neutralization of crystalline monocalcium phosphate obtained at an acid rate of 400%, the content of $P_2O_{5св.}$ decreases from 20.0% to 2.5% and from 19.8% to 2.6%, respectively, for monocalcium phosphate obtained at crystallization temperatures of 60 and 80°C.

Thus, in laboratory conditions the possibility of obtaining fodder, crystalline monocalcium phosphate was established, the optimal technological parameters of all stages of the process were determined: concentrations of EFC 45-55% P_2O_5, norm 350-500%, temperature - 100 ° C, duration of the process 180-300 min.

§ 3.7. Study of crystalline monocalcium phosphate extraction process

Similar tests were carried out for obtaining monocalcium phosphate (cooling rate was regulated starting from 65 C). The obtained results are given in Tables 3.15-3.16.

From the results obtained, it can be seen that the main factor affecting the filtration rate of monocalcium phosphate is the cooling rate of solutions. Decreasing the solution cooling rate from 15 to 5.0° C/hour at a solution concentration of 55% and process temperature of 60° C filtration time decreases from 1.68 min to 0.53 min. Decreasing the cooling rate of the monocalcium phosphate solution from 15 to 5.0° C/hour the sludge removal increases from 330 to 750 kg/m^3 -hour for a solution concentration of 60%.

Changing the concentration and temperature of the solution in the range of 55-60% and 60-65° C have no noticeable effect on the character of change in filtration time and sludge removal. Slow cooling rate favors the precipitation of fewer impurities. As the cooling rate decreases, the amount of SO_3, MgO, Al O_{23}, Fe O_{23} and F impurities in the product decreases by about half. Increasing the concentration of solutions above 55 % leads to an increase in the impurity content in the finished product.

Table 3.15.

Effect of solution concentration, temperature and cooling rate on technological parameters of crystallization and chemical composition of monocalcium phosphate

C, %	Tempe- The nature of it, °C	Speed cooling, °C/hour	Correspond ingly. stitching G:T.	Time filter-i. 200 ml, min	Take off sludge, kg/m^2 -h	Chemical composition of the product, wt. %						
						P_2O_5	SO_3	CaO	MgO	Al_2O_3	Fe_2O_3	F
Initial			-		-	50,01	0,768	4,366	1,709	1,332	0,860	0,171
55	60	15,0	7,72	1,63	330	54,20	0,190	18,70	0,067	0,096	0,085	0,014
		10,0	7,81	0,95	580	54,46	0,185	18,81	0,062	0,087	0,078	0,012
		5,0	7,86	0,53	750	54,52	0,182	18,84	0,060	0,081	0,074	0,009
	65	15,0	8,91	1,77	350	54,41	0,192	18,35	0,069	0,097	0,088	0,014
		10,0	9,05	1,04	610	54,62	0,183	18,27	0,064	0,089	0,081	0,011
		5,0	9,11	0,58	780	54,65	0,181	18,21	0,062	0,083	0,077	0,009
Initial			-	-	-	55,00	0,845	4,80	1,880	1,465	0,946	0,160
60	60	15,0	6,60	1,10	580	54,53	0,189	18,22	0,067	0,094	0,083	0,013
		10,0	6,68	0,67	710	54,71	0,181	18,10	0,062	0,086	0,078	0,011
		5,0	6,71	0,40	890	54,73	0,180	18,08	0,060	0,081	0,075	0,009
	65	15,0	7,61	1,00	640	54,88	0,187	18,15	0,065	0,092	0,080	0,012
		10,0	7,70	0,60	790	55,03	0,180	18,03	0,061	0,084	0,075	0,010
		5,0	7,74	0,35	980	55,06	0,179	17,99	0,059	0,080	0,074	0,008

Table 3.16

Degree of transition of solution components into product and chemical composition of mother liquor after crystallization of monocalcium phosphate

C, %	Tempe- The nature of it, °C	Speed cooling, °C/hour	Degree of transition of components into the product, %							Chemical composition of mother liquor, wt. %						
			$P\ O_{25}$	SO_3	CaO	MgO	$Al\ O_{23}$	$Fe\ O_{23}$	F	$P\ O_{25}$	SO_3	CaO	MgO	$Al\ O_{23}$	$Fe\ O_{23}$	F
55	60	15,0	18,24	4,76	67,64	0,80	0,16	1,96	0,87	31,61	0,498	1,166	1,105	0,87	0,56	0,707
		10,0	18,19	4,72	67,53	0,79	0,15	1,94	0,85	31,56	0,501	1,162	1,108	0,89	0,58	0,709
		5,0	18,17	4,71	67,51	0,78	0,14	1,93	0,82	31,54	0,502	1,159	1,110	0,90	0,59	0,712
	65	15,0	18,78	4,88	68,35	0,83	0,17	2,00	0,91	31,34	0,491	1,147	1,162	0,92	0,61	0,715
		10,0	18,71	4,84	68,21	0,82	0,16	1,98	0,89	31,28	0,494	1,142	1,167	0,95	0,63	0,717
		5,0	18,68	4,83	68,15	0,81	0,15	1,97	0,86	31,25	0,495	1,140	1,169	0,96	0,64	0,719
60	60	15,0	19,23	4,96	69,60	0,86	0,22	2,09	0,96	32,13	0,479	1,184	1,182	0,98	0,67	0,725
		10,0	19,17	4,93	69,47	0,83	0,21	2,04	0,94	32,18	0,474	1,181	1,187	1,00	0,70	0,729
		5,0	19,15	4,92	69,43	0,82	0,20	2,02	0,91	32,20	0,471	1,180	1,189	1,01	0,71	0,731
	65	15,0	19,86	5,04	70,25	0,91	0,27	2,15	0,99	32,02	0,473	1,178	1,179	1,04	0,73	0,733
		10,0	19,79	5,02	70,13	0,89	0,23	2,12	0,96	32,08	0,468	1,175	1,174	1,07	0,75	0,737
		5,0	19,77	5,01	70,12	0,88	0,21	2,11	0,94	32,10	0,465	1,174	1,172	1,08	0,76	0,740

The mother liquors after crystallization of monocalcium phosphate can be reused, after concentration, for crystallization of monocalcium phosphate.

Thus, the results of research show that the optimal parameters for obtaining feed monocalcium phosphate are the concentration of suspension of not less than 60%, cooling of suspension at a rate of not more than 5.0 ° C / h to a temperature of 60-65 ° C, and the resulting products fully meets the requirements of GOST 23999-80 for feed phosphates.

§ 3.8. Investigation of the multiplicity of mother liquor utilization

In order to increase the yield of P_2O_5 into the product and to obtain a product that meets the requirements of regulatory documentation, the influence of reuse of the mother liquor on the composition of the obtained crystalline monocalcium phosphate of feed and higher purity was studied. In this case, the mother liquor after crystallization of the product from the previous stage was subjected to evaporation to a given concentration (Tables 3.17 and 3.18).

During the experiments we used solutions subjected to concentration after their obtaining by neutralization of desulfurized and desulfurized solution of EFC from CK phosphate with calcium carbonate at a rate of 400 % of stoichiometry for the formation of monocalcium phosphate. In the experiments 200 g of 55 % solution of monocalcium phosphate with the temperature of 90 °C were poured into a water jacketed reactor cooled with water. After cooling to 60 °C, with constant stirring of the solution, the refrigerant supply was reduced, cooling at a rate of 5.0 °C/hour. When the temperature reached 60°C, cooling was stopped, the suspension was separated on a Buechner funnel (filtering surface 0.005 m).[2]

The crystals on the filter were pressed off by air suction for 2 minutes. The crystals on the filter were washed with acetone and dried at 100 C for 1 h.

Table 3.17.

Influence of mother liquor reuse on the composition of the obtained product

Stage	Sample	Chemical composition, wt. %						
		P O_{25}	SO_3	CaO	MgO	$AL_2 OH_3$	$Fe_2 O_3$	F
Initial EFC		17,0	0,230	1,58	0,490	0,38	0,25	0,310
1	Initial	50,0	0,676	4,647	1,440	1,118	0,735	0,909
	Crystals	54,20	0,190	18,70	0,067	0,096	0,085	0,009
	Solution	31,61	0,498	1,166	1,105	0,870	0,560	0,673
2	Initial	27,03	0,415	2,36	0,924	0,720	0,465	0,557
	Crystals	54,14	0,192	18,72	0,070	0,098	0,086	0,010
	Solution	31,64	0,570	0,955	1,313	1,028	0,655	0,788
3	Initial	27,01	0,464	2,358	1,069	0,830	0,530	0,637
	Crystals	54,11	0,195	18,75	0,073	0,099	0,088	0,012
	Solution	31,66	0,057	0,956	1,523	1,191	0,750	0,903
4	Initial	26,99	0,510	2,50	1,215	0,940	0,598	0,716
	Crystals	53,81	0,430	19,30	0,081	0,112	0,104	0,014
	Solution	31,755	0,679	1,09	1,732	1,34	0,844	1,013

Table 3.18

Effect of mother liquor reuse on process performance

Stage	Correspondingly. stitching G:T.	Time filtering, min.	Take off sludge, kg/m^2-h	Degree of transfer into the product, wt. %						
				P O_{25}	SO_3	CaO	MgO	$AL_2 OH_3$	$Fe_2 O_3$	F

1	7,72	0,78	589	18,17	4,71	67,51	0,78	0,14	1,93	1,33
2	7,72	0,71	561	18,16	4,15	71,72	0,66	0,12	1,65	1,14
3	7,72	0,68	552	18,15	3,71	71,70	0,57	0,10	1,45	1,00
4	7,71	1,25	420	17,99	7,58	69,64	0,60	1,07	1,57	1,39

From the experimental data (Tables 3.17 and 3.18) it can be seen that during the threefold reuse of the stock solution, the increase in the degree of transition of impurities in the product crystals is relatively low and is, %: for SO_3 - 4.19; MgO - 0.67; Al O_{23} - 0.12; Fe O_{23} - 1.68 and F - 0.82. This is due to the fact that during the reuse of the mother liquor, the impurity content of the product gradually increases. The yield of $P_2 O_5$ in qualified feed monocalcium phosphate is 67.51%.

Thus, threefold reuse of the recycled mother liquor can be considered optimal, which ensures the production of fodder monocalcium phosphate that meets the requirements of GOST 23999-80.

§ 3.9. Study of rheological properties of crystalline monocalcium phosphate solutions

Investigations of rheological properties of products during calcium carbonate decomposition at increased EFC rate. The EFC rate was varied in the range from 300 to 450%, and the temperature from 40 to 100°C. Experimental data are presented in Tables 3.19 and 3.20.

Table 3.19 summarizes the rheological properties of the monocalcium phosphate suspension.

Table 3.19.

Effect of the rate of extraction phosphoric acid and temperature on the density and viscosity of the suspension formed during the decomposition of calcium carbonate at an increased rate of acid

№	Norms EFC, %	Density, g/cm³					Viscosity, mPa.s				
		40°C	60°C	80°C	90°C	100°C	40°C	60°C	80°C	90°C	100°C
1	450	1,263	1,252	1,244	1,241	1,239	3,234	2,258	1,487	1,256	1,245

2	400	1,320	1,308	1,300	1,297	1,295	3,593	2,509	1,652	1,395	1,379
3	375	1,379	1,367	1,358	1,355	1,352	6,314	4,243	2,904	2,358	1,344
4	350	1,434	1,421	1,412	1,409	1,406	9,035	5,977	4,156	3,322	3,309
5	325	1,480	1,467	1,458	1,455	1,452	12,055	7,975	5,546	4,432	4,415
6	300	1,527	1,513	1,504	1,500	1,497	17,753	11,746	8,167	6,527	6,502

As the EFC rate of the suspension decreases, the density and viscosity increase slightly and are 1.263-1.527 g/cm^3 and 3.234-17.753 mPa-s at 40° C.

Increasing the temperature of the suspension causes the density and viscosity of the suspension to decrease. At an EFC rate of 450% increasing the temperature from 40 to 100° C, the density decreases from 1.263 g/cm^3 to 1.239 g/cm^3 , and the viscosity under these conditions decreases from 3.234 mPa-s to 1.245 mPa-s. This indicates acceptable rheological properties of the monocalcium phosphate suspension.

Table 3.20 shows the rheological properties of the mother liquor after filtration of crystalline monocalcium phosphate suspension.

Table 3.20.

Effect of the rate of extraction phosphoric acid and temperature on the density and viscosity of the suspension formed during the decomposition of calcium carbonate at an increased rate of acid

№	Norms EFC, %	Density, g/cm^3					Viscosity, mPa·s				
		40°C	60°C	80°C	90°C	100°C	40°C	60°C	80°C	90°C	100°C
1	450	1,175	1,164	1,157	1,154	1,151	3,008	2,099	1,383	1,268	1,206
2	400	1,228	1,217	1,209	1,206	1,203	3,341	2,333	1,536	1,397	1,327
3	375	1,282	1,271	1,263	1,260	1,257	5,872	3,946	2,701	2,193	2,084
4	350	1,334	1,322	1,313	1,310	1,307	8,402	5,559	3,865	3,089	2,935
5	325	1,376	1,364	1,356	1,353	1,349	11,211	7,417	5,158	4,122	3,875
6	300	1,420	1,407	1,399	1,395	1,392	16,510	10,924	7,595	6,069	5,644

As the EFC rate of the suspension decreases, the density and viscosity increase slightly and are 1.175-1.420 g/cm^3 and 3.008-16.510 mPa-s at 40° C.

Increasing the temperature of the suspension causes the density and viscosity of the suspension to decrease. At an EFC rate of 450% increasing the temperature from 40 to 100° C, the density decreases from 1.175 g/cm^3 to 1.151 g/cm^3 , and the viscosity under these conditions decreases from 3.008 mPa-s to 1.206 mPa-s. This indicates acceptable rheological properties of the monocalcium phosphate suspension.

According to the results of the conducted researches it is established that within all studied technological parameters the values of density and viscosity of monocalcium phosphate suspension and mother liquors obtained after filtration have high fluidity and allow their transportation by pumping devices without any restrictions.

§ 3.1.0. Material balance of crystalline monocalcium phosphate production

Material balance of crystalline monocalcium phosphate production by cyclic methods at increased rate of phosphoric acid is shown in Figure 3.11.

In the first step, 17544 kg of purified EFC is evaporated to 50% P_2O_5 . Small crystals of monocalcium phosphate are formed in the solution. To improve the filtration process, the evaporated EFC is kept under stirring for 3 hours and filtered. Crystals of monocalcium phosphate in the amount of 1403 kg, formed due to the content of CaO in the initial acid, washed with water in the amount of 2807 kg, dried. and obtained 1000 kg of crystalline monocalcium phosphate. The filtrate is mixed with wash water and fed to the second stage for limestone decomposition.

For this purpose 7720 kg of mother liquor from the first stage are mixed with 3200 kg of purified EFC and 220 kg of limestone. The mixture is evaporated to the content of 50% P_2O_5 , crystallized and 5695 kg of suspension of monocalcium phosphate in EFC is fed to the filter. The precipitate of 1403 kg is washed with 2807 kg of water, dried and 1000 kg of crystalline monocalcium phosphate is obtained. Washing water and filtrate in the form of mother liquor are fed to the decomposition of a new portion of limestone.

Fig. 3.11. Material balance of crystalline monocalcium phosphate production

The optimal norms of technological mode of production of feed monocalcium phosphate on the basis of EFC from phosphate rock from CK phosphorites were established, which are shown in Table 3.22.

Table 3.22

Norms of technological mode of production of granulated and crystalline feed monocalcium phosphate

№	Name of parameters	Paramete r values
Crystalline feed monocalcium phosphate		
1.	EFK concentration, % P_2O_5	45-55
2.	EFC norm, %	300-500
3.	Decomposition process temperature, °C	80-100
4.	Slurry residence time in the reactor, min	180-300
Slurry evaporation and crystallization of monocalcium phosphate		
5.	Residual water content in pulp, %	40-50
6.	Temperature of packing process, °C	100-105
7.	Crystallization temperature of monocalcium phosphate, °C	60-80
8.	L:T ratio in pulp	2,5-3,5
Filtration and washing of monocalcium phosphate crystals		
9.	Filtration vacuum, kgf/cm^2	0,45-0,65
10.	Water consumption for sludge washing, t/t dry sludge	0,4-0,7
11.	Humidity of monocalcium phosphate crystals, %	10-15
12.	Density of mother liquor, g/cm ^{3}	1250-1400
Neutralization and drying of monocalcium phosphate crystals		
13.	Calcium carbonate consumption for neutralization of free acidity, t/t dry sludge	0,1-0,15
14.	Drying temperature, °C	100-110
15.	Drying process time, min	10-20

Thus, the optimal technological parameters for obtaining crystalline monocalcium phosphate by decomposition of limestone purified from fluorine and sulfates 45-55% P_2O_5 EFC from phosphorites CK have been established: decomposition temperature - 100°C, acid rate - 400%, process duration 180-300 min., at which the product is obtained with the content - 54,2% P_2O_5 ; 18,7% CaO; 0,012% F.

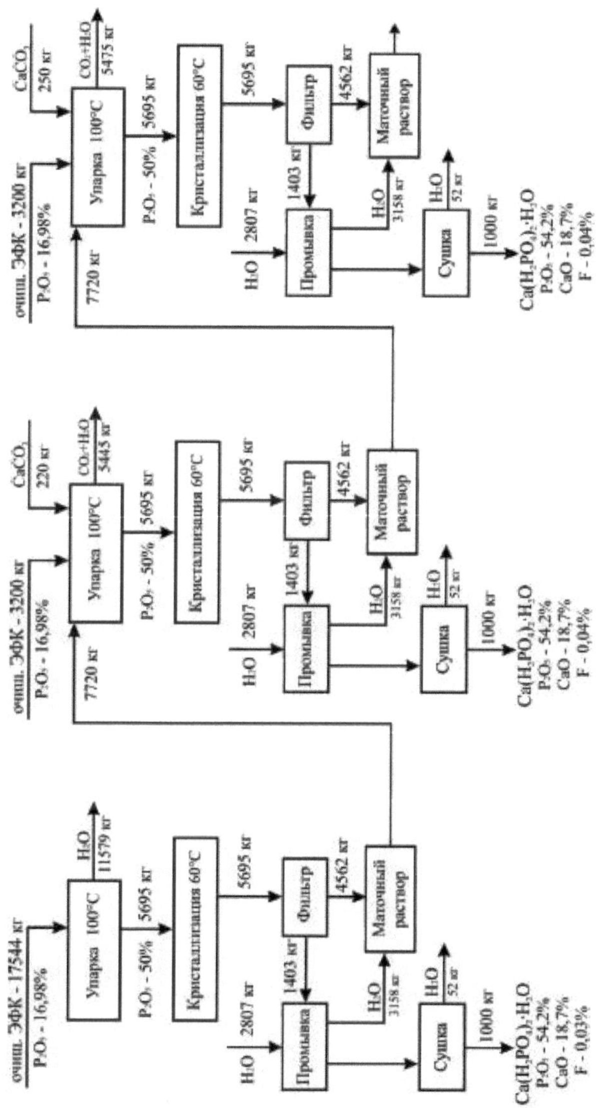

§ 3.11. Physicochemical studies of the obtained crystalline monocalcium phosphate

To conduct research by physicochemical methods of analysis obtained crystalline monocalcium phosphate by decomposition of limestone with concentrated EFC at elevated rate. It is shown that at the concentration of purified EFC 45-55% $P_2 O_5$ and the rate of 400% obtained monocalcium phosphate with a content of 54-55% $P_2 O_5$ and 0.009-0.02% fluorine.

The X-ray diffraction pattern (Figure 3.12) shows only diffraction maxima characteristic of monocalcium phosphate with interplanar distances of 11.75, 4.93, 3.007, 2.95 Å.

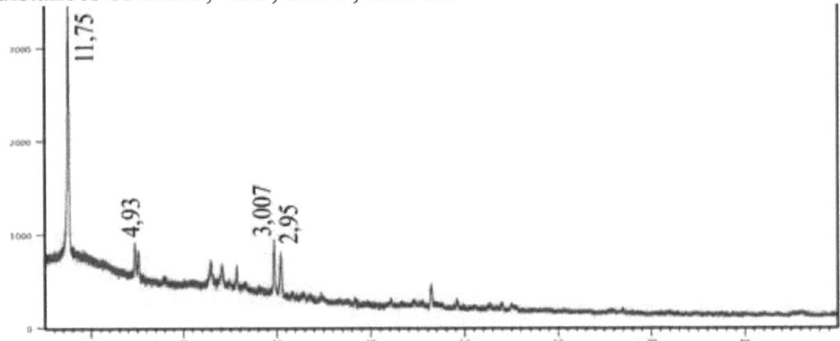

Fig. 3.13. X-ray diagram of crystalline monocalcium phosphate

On the IR - spectrum (Fig. 3.14) there are frequencies of vibrations characterizing vibrations related to RO_4 440,54-1077,68 cm^{-1} and crystalline water - 1647,11-2897,93 $cm.^{-1}$

Electron-microscopic image of monocalcium phosphate crystals obtained after separation from the mother liquor, as well as the results of their elemental chemical analysis are shown in Figure 3.15 and Table 3.22. Scanning microscopic analysis of crystalline monocalcium phosphate shows the following content of elements of composition: O-51,28%, F-0,11%; Na-1,51%; Mg-1,89%; Al-0,92%; Si-0,11%; P-23,38%; S-0,31%, which corresponds to their content in feed monocalcium phosphate.

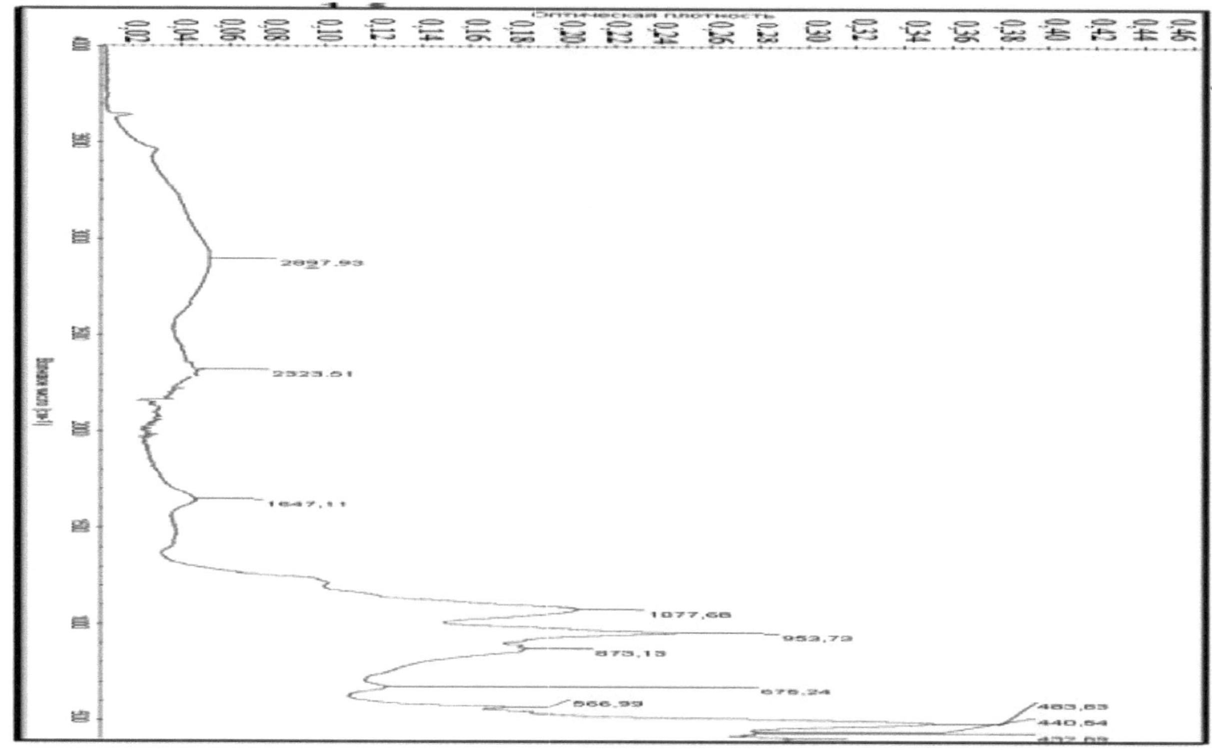

2927,93
2323,51
1647,11
1077,66
953,73
873,15
678,24
566,93
483,63
440,64

Fig. 3.14. IR spectrum of feed monocalcium phosphate

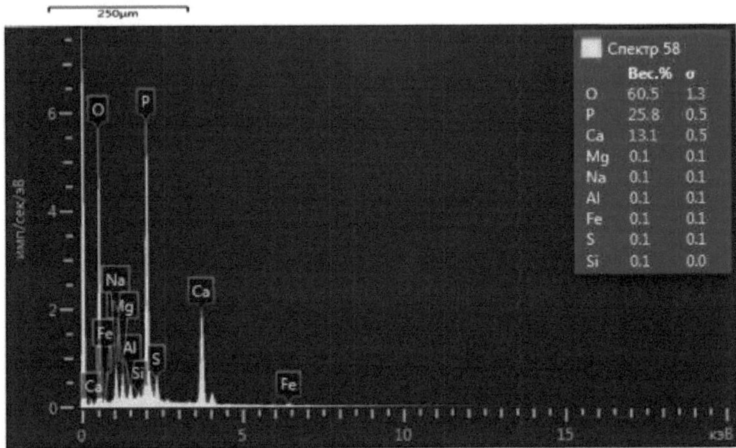

Fig. 3.15. Scanning microscopic analysis of feed monocalcium phosphate

Talitsa 3.23

Results of elemental chemical analysis of monocalcium phosphate

Element	Weight. %	Sigma, Wt. %
O	60.51	1.29
Na	0.09	0.03
Mg	0.06	0.03

Al	0.08	0.04
Si	0.06	0.02
P	25.8	0.22
S	0.09	0.04
Ca	13.08	0.35
Fe	0.07	0.04
Summation:	100.00	

Thus, the possibility of obtaining crystalline monocalcium phosphate has been experimentally established, the optimal parameters of all stages of the process have been determined, its physical, chemical and commercial properties have been clarified.

§ 3.12. Approbation of technology for production of monocalcium phosphate on the basis of evaporated extraction phosphoric acid and calcium carbonate

On the basis of systematic experimental studies two variants of production of desulfurized feed monocalcium phosphate on the basis of EFC from CK phosphorites are proposed.

Technology of obtaining granulated feed monocalcium phosphate. The technological process for the first variant of obtaining granulated feed monocalcium phosphate on the basis of evaporated EFC from CK phosphate rock includes:

- evaporation of defluorinated and desulfurized EFC to a concentration of 45-55% $P_2 O$;$_5$

- decomposition of calcium carbonate by evaporated, partially purified from fluorine and sulfates EFC;

- Granulation and drying in the presence of retort of defluorinated feed monocalcium phosphate.

The essence of the technology consists in the decomposition of calcium carbonate with defluorinated, desulfurized and evaporated EFC with concentration of $P_2 O_5$ 45-55% at a rate of 95-100% of stoichiometry, temperature 90-100° C and duration of the process 20-40 min. Meanwhile, granulation and drying are carried out at a temperature of 105-110° C using a drum granulator (BG) and drum dryer (BD) apparatus in the presence of retort.

The obtained desulfurized feed granular monocalcium phosphate has the composition (wt. %): $P_2 O_{5общ}$. - 52,18; $P_2 O_{5усв}$. - 52,01; $P_2 O_{5водн}$. - 51,78; CaO - 27,14; F - 0,16.

Technology of obtaining crystalline feed monocalcium phosphate. The technological process for the second variant of obtaining feed crystalline monocalcium phosphate on the basis of EFC obtained from phosphate rock of CK includes:

- decomposition of calcium carbonate at high rates of EFC evaporated to the concentration of 45-55% $P_2 O_5$ and purified from fluorine and sulfates;
- Crystallization and separation of monocalcium phosphate crystals;
- Drying and packaging of crystalline monocalcium phosphate;
- return of the mother liquor to the calcium carbonate decomposition stage.

Calcium carbonate was decomposed with purified EFC at a rate of 400% of the stoichiometrically necessary amount at a temperature of 90-100° C, with constant stirring for 180 - 300 minutes, separated insoluble residue, evaporated the pulp to precipitate crystals of monocalcium phosphate, cooled to a temperature of 30° C, separated crystals of monocalcium phosphate and neutralized with calcium carbonate. Monocalcium phosphate of composition (wt. %) was obtained: $P_2 O_{5общ}$.-55.67; $P_2 O_{5усв}$.-55.40; $P_2 O_{5водн}$.-55.18; CaO-26.81; F-0.012.

The prototypes of granular and crystalline feed monocalcium phosphate, in which fluorine content is not more than 0.18% and 0.017%, respectively, have been developed.

Physicochemical and commercial properties of desulfurized, feed monocalcium phosphate were studied. The obtained samples of feed monocalcium phosphate meet the requirements for feed phosphates GOST - 23999-80 and were submitted for testing (Table 3.24).

Table 3.24.

Chemical and physicochemical parameters and comparative characterization of the obtained monocalcium phosphate

№ n/a	Name of indicators	Norma		
		According to GOST 23999-80	In fact	
			I-option	II-option

1	Mass fraction of total phosphate, (P_2O_5), 1st grade, %, not less 2nd grade, %, not less	55-56 50-51	- 52,18	55,67 -
2	Mass fraction of calcium, %, not less than	18	18,65	19,15
3	Hydrogen ion activity index unit pH, not less than	3	3,26	3,20
4	Mass fraction of water, %, max.	4	2,67	1,12
5	Mass fraction of fluorine, %, max.	0,2	0,18	0,07
6	Mass fraction of arsenic, %, not more than	0,005	otc.	otc.
7	Mass fraction of lead, %, not more than	0,002	otc.	otc.
8	Mass fraction of ash insoluble in hydrochloric acid, %, not more than	10	8	6

Thus, the conducted tests have shown adequate reproducibility of the results of laboratory studies, concerning technological parameters of realization of the processes of obtaining desulfurized monocalcium phosphate and the quality of the obtained products. The technologies, if equipped with appropriate equipment, will provide economic efficiency.

§ 3.13. Technical and economic calculations of feasibility of feed monocalcium phosphate production

Production of feed, granular and crystalline monocalcium phosphates on the basis of EPC obtained from phosphate rock of CK includes the following stages:

Granulated monocalcium phosphate: purification of EPC from fluorine and sulfate, acid stripping, decomposition of limestone with decomposed phosphoric acid at a concentration of 45-55% P_2O_5 at stoichiometric rate; granulation in the presence of retort and drying.

Crystalline monocalcium phosphate: purification of EFC from fluorine and sulfates, acid evaporation, decomposition of limestone with evaporated phosphoric acid at high (350-450%) rates; cooling, crystallization and separation of monocalcium phosphate crystals; return of stock solutions to the initial stage of the process; washing and drying of crystalline monocalcium phosphate.

The costs of monocalcium phosphate production consist of the cost of EPC, liquid glass, soda ash, washed, MOFK to obtain purified EPC from fluorine and sulfate, limestone, as well as processing, marketing and other costs.

Prices for 1 ton of raw materials and energy resources are accepted: EFC with 17% P_2O_5 - 239480 soums, MOFC - 490539 soums (according to JSC "Ammophos-Maksam"), soda ash - 2076,824 soums (RTSB), limestone - 500 thousand soums, liquid glass 50% solution - 1500 thousand soums.

Prices for energy resources are taken 411 UZS per 1 kWh of electricity, steam 148596 UZS per 1 Gcal, recycled water 434477 per 1000 m^3, compressed air 92790 UZS per 1000 m^3.

Production of granulated feed monocalcium phosphate. To obtain 1 ton of feed monocalcium phosphate it is necessary 2.848 tons of EFC with the content of 17% P_2O_5. For partial purification of acid from fluorine and sulfates 0.139 tons of soda ash, 0.07 tons of liquid glass, 0.210 tons of MOFC are consumed.

The cost of acid purification will be:

2,848 x 239,480 = 682,039 thousand soum cost of EFC.

0,139 x 2076,824 = 288,678 thousand soums cost of soda ash.

0.070 x 1500 = 105.00 thousand soums cost of liquid glass.

0,210 x 490,539 = 103,013 thousand soums cost of MOFC.

The total costs of desulfurization and desulfurization will be:

682,039 + 288,678 + 105,00 + 103,013 = 1178,73 thousand soums.

To produce 1 ton of finished feed monocalcium phosphate, another 0.350 tons of limestone is needed at a cost of:

0.350 x 500.00 =175.00 thousand soums.

Total raw material costs will be:

1178,73 + 175,00 = 1353,73 thousand soums.

To produce 1 ton of monocalcium phosphate, 250 kW of electricity, 10.44 GJ or 2.5 Kcal of steam and 25 m^3 of compressed air are consumed. Energy costs of monocalcium phosphate production will be as follows:

2,50 x 148,596 = 371,490 thousand soums cost of steam.

25 x 0.09279 = 2.320 thousand UZS cost of compressed air.

66,7 x 0,380 =25,335 thousand soums cost of natural gas.
250 x0.411 = 102,750 thousand soums cost of electricity.

Total energy consumption for the production of 1 ton of monocalcium phosphate will be as follows:

371,490 + 2,320 + 25,335 + 102,750 = UZS 501,895 thousand.

The total cost of raw materials and energy resources will be:

1353,73 + 501,895 = 1855,625 thousand soums.

Technological expenses for monocalcium phosphate pulp production, evaporation and drying are assumed to be 40% of the cost of raw materials and energy resources:

1855,625 x 0,40 = 742,25 thousand soums.

Technological cost of monocalcium phosphate will be as follows:

1855,625 + 742,25 = 2597,875 thousand soums.

Expenses of the period and sales of monocalcium phosphate are assumed to be 10% of the technological cost:

2597,875 x 0,1 = 259,788 thousand soums.

The full factory cost of production will be:

2597,875 + 259,788 = 2857,663 thousand soums.

The wholesale price of feed monocalcium phosphate is 850 USD. The wholesale price of feed monocalcium phosphate is 850 USD without transportation and customs costs, and taking into account transportation and customs costs we take 1000 USD on average. USD or 1000 x 10855=10,855.0 thousand soums, where 10,855 soums dollar exchange rate in soums as of 20.12.2021.

Savings from each ton of feed monocalcium phosphate will be as compared to imported monocalcium phosphate,

10,855.0 - 2857.663 = 7997.337 thousand soums.

When producing 10 thousand tons of granulated feed monocalcium phosphate, the savings will amount to 79,973.37 million soums.

Production of crystalline feed monocalcium phosphate. To obtain 1 ton of monocalcium phosphate it is necessary 3,200 tons of EFC with the content of 17% P_2O_5 . For partial purification of acid from fluorine and sulfates 0.156 tons of soda ash, 0.079 tons of liquid glass, 0.240 tons of MOFC are consumed.

The cost of acid purification will be:

3,200 x 239,480 = 766,3361 thousand soum cost of EFC.
0.079 x 1500 = 118.5 thousand soums cost of liquid glass.
0,240 x 490,539 = 117,730 thousand soums cost of MOFC.
0,156 x 2076,824 = 323,984 thousand soums cost of soda ash.

The total costs for EFC desulfurization and desulfurization will be as follows:

766,336 + 118,5 + 117,730 + 323,984 = 1326,55 thousand soums.

To produce monocalcium phosphate, an additional 0.220 tons of limestone is needed at a cost of:

0.220 x 500.00 = 110.00 thousand soums.

Total raw material costs will be as follows

1326,55 + 110,00 = 1436,55 thousand soums.

To produce 1 ton of crystalline feed monocalcium phosphate 285 kW of electricity, 2.85 Kcal of steam, 7.720 m^3 of recycled water and 28.5 m^3 of compressed air are consumed. Energy costs of crystalline monocalcium phosphate production will be:

7,720 x 0,435 = 3,358 thousand soums cost of recycled water.
2.85 x 0.1486 = 423.499 thousand soums cost of steam.
28,5 x 0,0928 = 2,645 thousand soums cost of compressed air.
76,04 x 0,380 = 28,895 thousand soums cost of compressed air.
855 x 0.411 = 351.405 thousand soum cost of electricity.

Total expenditures of energy resources for production of 1 ton of crystalline monocalcium phosphate will be as follows:

3,358 + 423,499 + 2,645 + 28,895 + 351,405 = 809,802 thousand soums.

Expenses for raw materials and energy resources will amount to:

1436,55 + 809,802 = 2246,352 thousand soums.

Expenses for obtaining suspension and crystals of monocalcium phosphate, evaporation and drying are assumed to be 40% of the cost of raw materials and energy resources:

2246,352 x 0,40 = 898,541 thousand soums.

The technological cost of crystalline feed monocalcium phosphate will be:

2246,352 + 898,541 = 3144,893 thousand soums.

Expenses of the period and sales of crystalline monocalcium phosphate are taken 10% of the technological cost:

3144,893 x 0,1 = 314,489 thousand soums.

The full factory cost of production will be:

3144,893 + 314,489 = 3459,382 thousand soums.

The wholesale price of feed monocalcium phosphate is 850 USD. The wholesale price of feed monocalcium phosphate is 850 USD without transportation and customs costs, and taking into account transportation and customs costs we take 1000 USD on average. USD or 1000 x 10,855

= 10,855.0 thousand soums, where 10,855 soums dollar exchange rate in soums on 20.12.2021.

Savings from each ton of monocalcium phosphate will be as compared to imported monocalcium phosphate,

10,855.0 - 3,459.382 = 7,395.618 thousand soums.

When producing 10 thousand tons of crystalline feed monocalcium phosphate, the savings will amount to 73,956.18 million soums.

Thus, the conducted technical and economic calculations indicate a good profitability of monocalcium phosphate production on the basis of desulfated, desulfurized and evaporated solutions of EPC obtained from phosphate rock, high efficiency of production, allowing to obtain import-substituting chemical products, create additional jobs and meets all the requirements for feed monocalcium phosphate.

§ Conclusions to Chapter 3

To produce feed monocalcium phosphate from limestone and desulfurized, desulfurized EFC, the effect of temperature and process duration on the degree of limestone decomposition was investigated at an acid rate of 100% and a concentration of 17-50% P_2O_5

Increasing the temperature of the decomposition process from 20 to 80°C significantly increases the degree of limestone decomposition for all values of the process duration.

The optimal technological parameters of limestone decomposition have been established: decomposition temperature - 80-100°C, process duration - 30-60 min, particle size - 0.1-1.0 mm. Increasing the duration of the decomposition process and reducing the diameter of limestone particles also contributes to increase the degree of decomposition.

Increasing the concentration of EFC contributes to the reduction of fluorine content in the finished product, respectively, defluorinated and fertilizer monocalcium phosphate. However, this is not sufficient to obtain monocalcium phosphate of feed purity.

The studies have shown the possibility of obtaining defluorinated, fertilizer monocalcium phosphate on the basis of defluorinated and desulfurized EPC from phosphate rock of CK. Monocalcium phosphate obtained at EFC concentration up to 40 % P_2O_5 contains 51.97-54.99 % P_2O_5 , 25.63-28.24 % CaO. The fluorine content is 0.24-1.02%. The higher the concentration of initial EFC, the lower the fluorine content of monocalcium phosphate. To obtain monocalcium phosphate of feed purity it is necessary to carry out deeper purification of EPC from phosphorites of CC from fluorine.

The conducted studies have shown the possibility of obtaining feed monocalcium phosphate on the basis of desulfurized and desulfurized EPC from phosphate rock of CK. Monocalcium phosphate obtained at EFC concentration from 45 to 60 % P_2O_5 contains 52,65-55,26 % P_2O_5, 26,55-28,33 % CaO. The fluorine content is 0.12-0.19%. The higher the concentration of initial phosphoric acid, the lower the fluorine content of feed monocalcium phosphate. The obtained samples of feed monocalcium phosphate meet the requirements for feed phosphates GOST - 23999-80.

Drying of monocalcium phosphate with high moisture content is not economically justified. Therefore, in order to reduce the moisture content of the product fed for drying and granulation, the effect of the ratio of MCP:retort on the change in the chemical composition and moisture content of the product was studied.

On the basis of the obtained results the technological scheme and material balance of fertilizer and fodder monocalcium phosphate production from desulfurized, desulfurized and evaporated EFC were developed.

In order to obtain desulfurized monocalcium phosphate in crystalline form, without extraneous impurities, the process of decomposition of limestone by desulfurized and desulfurized EFC from phosphate rock of CK, previously evaporated to the content of 40-55% P_2O_5 at its rate of 300-500% of stoichiometry for the formation of monocalcium phosphate was investigated.

Phosphoric acid mass of decomposition of limestone 45% by P_2O_5 phosphoric acid at rates of 300 and 400% of stoichiometry is practically not filtered. A similar pattern is observed at a rate of 55% phosphoric acid 300%. The best filtration results are observed when 50% P_2O_5 phosphoric acid is used at rates of 300-500% and when 55% P_2O_5 phosphoric acid is used at 500%. In this case, the sludge removal of monocalcium phosphate is 330-450 $kg/m^2 \cdot h$ and the $P_2O_{5общ.}$ content is 53.6-54.8%, $P_2O_{5св.}$ 11.2 - 14.5%, CaO 16.6 - 17.5%.

In laboratory conditions the possibility of obtaining fodder, crystalline monocalcium phosphate was established, optimal technological parameters of all stages of the process were determined: EFC concentrations 45-55% P_2O_5, norm 350-500%, temperature - 100°C, process duration 180-300 min.

The results of research show that the optimal parameters for obtaining fodder monocalcium phosphate are the concentration of suspension not less than 60%, cooling of suspension at a rate of not more than 5.0 ° C / h to a temperature of 60-65 ° C, and the resulting products fully meets the requirements of GOST 23999-80 for fodder phosphates.

It was found that at threefold reuse of the stock solution the increase in the degree of transition of impurities into crystals of the product is relatively low and is, %: for SO_3 - 4.19; MgO - 0.67; Al O_{23} - 0.12; Fe O_{23} - 1.68 and F - 0.82. This is due to the fact that during the reuse of the mother liquor, the impurity content of the product gradually increases. The yield of P_2 O_5 in qualified feed monocalcium phosphate is 67.51%.

The possibility of obtaining fertilizer and fodder monocalcium phosphate was established, optimal parameters of all stages of the process were determined, physicochemical and commercial properties were established.

The basic technological scheme and material balance of production of fertilizer and feed monocalcium phosphate from defluorinated, desulfurized and evaporated EFC have been developed. Technical and economic calculations indicate the economic feasibility of production.

Principle technological schemes were developed and material balance of production of fertilizer and feed monocalcium phosphates from defluorinated, desulfurized and evaporated EFC was drawn up. Technical and economic calculations indicate the economic feasibility of production.

CHAPTER IV. STUDY OF PROCESSES OF MONO-POTASSIUM PHOSPHATE PRODUCTION ON THE BASIS OF MONOSODIUM PHOSPHATE AND POTASSIUM CHLORIDE

§ 4.1 Analysis of the system NaH_2PO_4-KH_2PO_4-NaCl-KCl-H_2O as applied to the production of mono-potassium phosphate

To select the optimal conditions for conversion of monosodium phosphate with potassium chloride and obtaining potassium dihydrophosphate, comparative studies of solubility in the four-component water-salt system K^+, Na^+ // Cl^-, $H_2PO_4^-$ - H_2O obtained at 25 and 100°C were carried out. The solubility diagram data of the four-component system in the form of a trapezoid, and in the form of a square at temperatures of 25 and 100°C are shown in Figures 4.1-4.3.

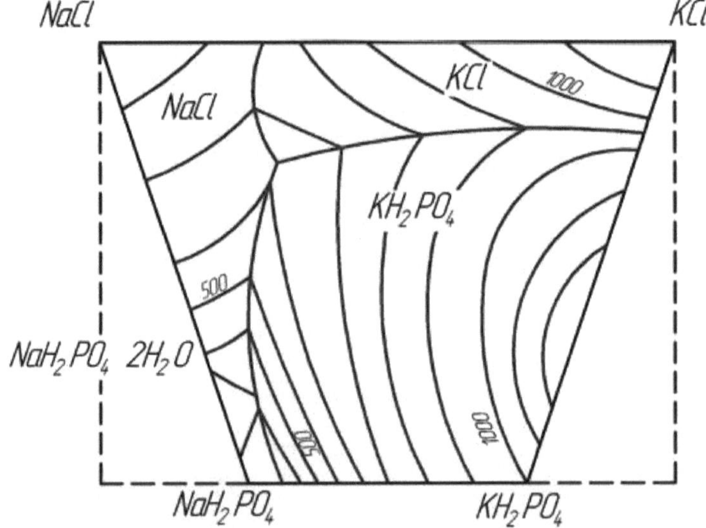

Fig. 4.1. Solubility diagram of the system K^+, Na^+ // Cl^-, $H_2PO_4^-$ - H_2O at 25°C.

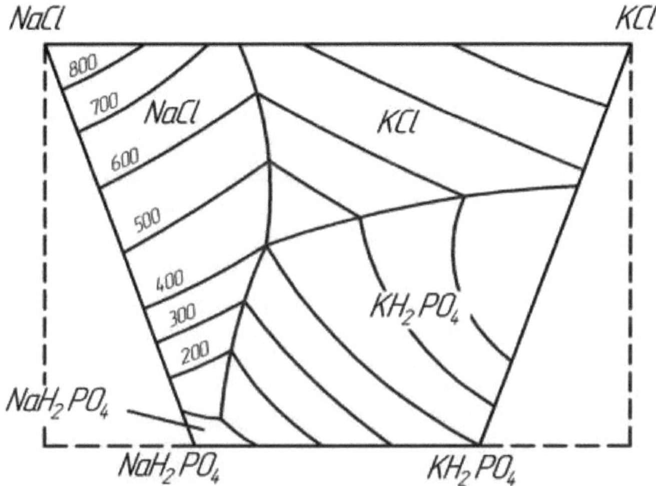

Fig. 4.2. Solubility diagram of the system K^+, Na^+ // Cl^-, $H_2PO_4^-$ - H_2O at 100°C.

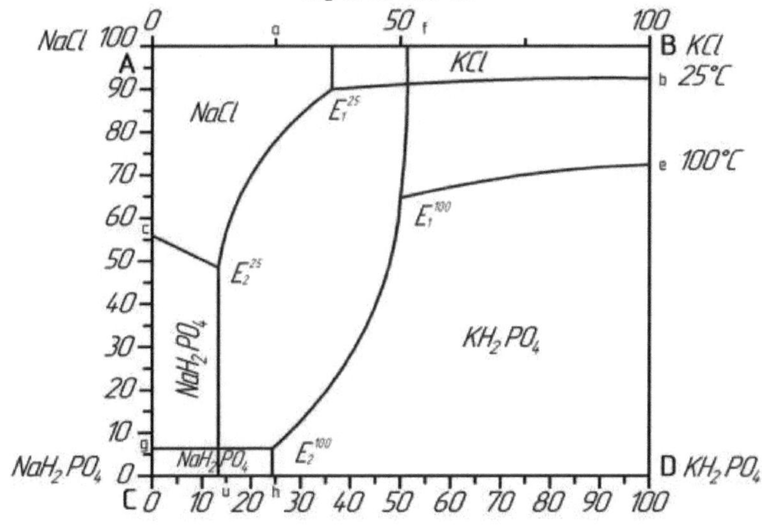

Fig. 4.3. Solubility diagrams of the systems K^+, Na^+ // Cl^-, $H_2PO_4^-$ - H_2O at 25 and 100°C.

The solubility diagram shows four crystallization fields of NaCl, KCl, $NaH_2 PO_4$ and $CH_2 PO_4$. E_1 and E_2 are eutonic points at temperatures of 25 and 100°C where the solution is in equilibrium with the three salts. These points are located outside the composition triangle in the solid phase of the salts and are incongruently saturated.

The crystallization field of $KN_2 PO_4$ on the solubility diagram of the system K^+ , Na^+ // Cl^- , $H_2 PO_4^-$ - $H_2 O$ at 25°C occupies a larger part of the square area than at 100°C. When considering the isothermal solubility diagram, it can be seen that as the temperature increases, the crystallization fields of $KN_2 PO_4$ decrease sharply, hence their solubility increases significantly. The saturation region of $CH_2 PO_4$ at 100°C is several times smaller than at 25°C.

The crystallization field of NaCl at 100°C is significantly increased, compared to the crystallization field at 25°C, due to a decrease in the field of $CH_2 PO_4$. Consequently, the solubility of $KN_2 PO_4$ is increased several times and such a solution is capable of precipitating this salt on cooling.

The CH $system_2$ PO_4 -NaH_2 PO PO_4 -NaCl- KCl-H_2 O has been used to analyze the phase composition of industrial processes.

Before starting to analyze the cyclic processes of potassium dihydrophosphate production, we have conducted preliminary studies of potassium dihydrophosphate production at different temperatures. As a result of this analysis, in our opinion, the most favorable conditions are provided by the process of obtaining potassium dihydrophosphate with sodium chloride release at 100°C and crystallization of potassium dihydrophosphate with cooling the suspension to 25°C.

On the basis of phase diagrams of solubility of mutual salt system K^+ , Na^+ // Cl^- , $H_2 PO_4^-$ -$H_2 O$, comparison of a number of technological cycles the scheme on the cycle Q_3 -M_3 -P_3 -S_3 -Q_3 was chosen (Fig. 4.4).

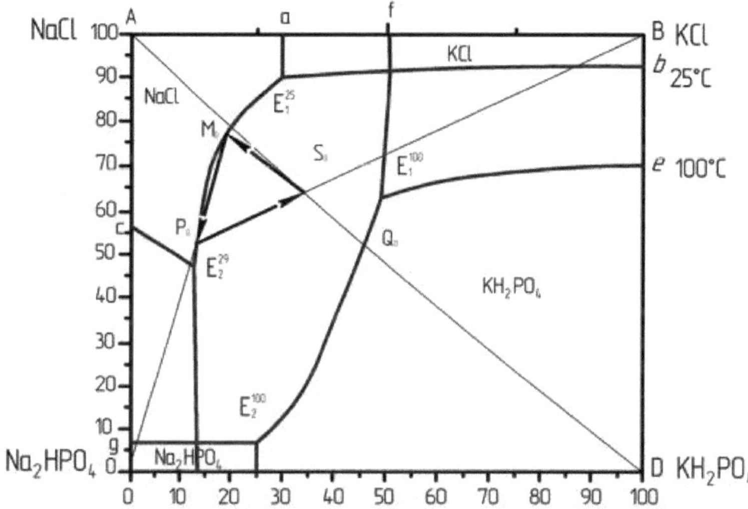

Figure 4.4. Diagram of potassium dihydrophosphate production.

In the first embodiment, sodium dihydrophosphate, which is a semi-product obtained by neutralizing phosphoric acid with sodium carbonate and potassium chloride, was used as a starting component.

The technological process proceeds as follows (Fig. 4.4): sodium dihydrophosphate is added to the circulating solution at point P_0 , potassium chloride is added at point S_0 , then the solution is evaporated to the saturation point Q_0 , lying on a line drawn from point D to point M_0 , which refers to the saturation point of sodium chloride at 100°C. After the precipitate has been separated, the mother liquor is cooled to a temperature of 25°C to point M_1 . At the temperature reached, the product is separated by filtration, then dried at 115-120°C. The remaining mother liquor is sent to the cycle. Thus, the process is closed.

In this case, the crystallization of KN_2 PO_4 will take place from a solution with a lower chloride content, allowing the production of KN_2 PO_4 with lower NaCl and relatively.

§ 4.2 Study of conversion of monosodium phosphate solutions by potassium chloride

The authors studied the solubility diagrams for the theoretical justification of obtaining potassium dihydrophosphate suitable for food production by conversion method from sodium dihydrophosphate obtained by neutralization of thermal phosphoric acid with soda ash and chemically pure potassium chloride.

To select the optimal conditions for conversion of monosodium phosphate solution obtained on the basis of EPC from CK phosphorites by potassium chloride flotation to obtain potassium dihydrophosphate, the previously studied systems were analyzed and comparative studies on solubility in the four-component water-salt system K^+ , Na^+ //Cl^- , $H_2 PO_4^-$ -H_2 O at 25 and 100°C were carried out in relation to the proposed compositions of the feedstock.

Analysis of NaH system$_2$ PO_4 -KH_2 PO_4 -NaCl-KCl-H_2 O showed the possibility of obtaining monokaliy phosphate by conversion of monosodium phosphate solution by flotation potassium chloride with preliminary release of sodium chloride. Therefore, further studies were aimed at obtaining monokaliy phosphate by conversion of monosodium phosphate solution with flotation potassium chloride.

Purified solutions of monosodium phosphate and flotation potassium chloride produced by UE "Dehkanabad Potash Plant" were used as initial components.

The technological process proceeds as follows (Fig. 1): sodium dihydrophosphate is added to the circulating solution at point P_0 , then potassium chloride is added at point S_0 , then the solution is evaporated to the saturation point Q_0 , lying on a line drawn from point D to point M_0 , which refers to the saturation point of sodium chloride at 100°C.

After the precipitate has been separated, the mother liquor is cooled to a temperature of 25°C to point M_1 , at the temperature reached, the product is separated by filtration, then dried at 115-120°C, the remaining mother liquor is sent to the cycle. Thus, the process is closed.

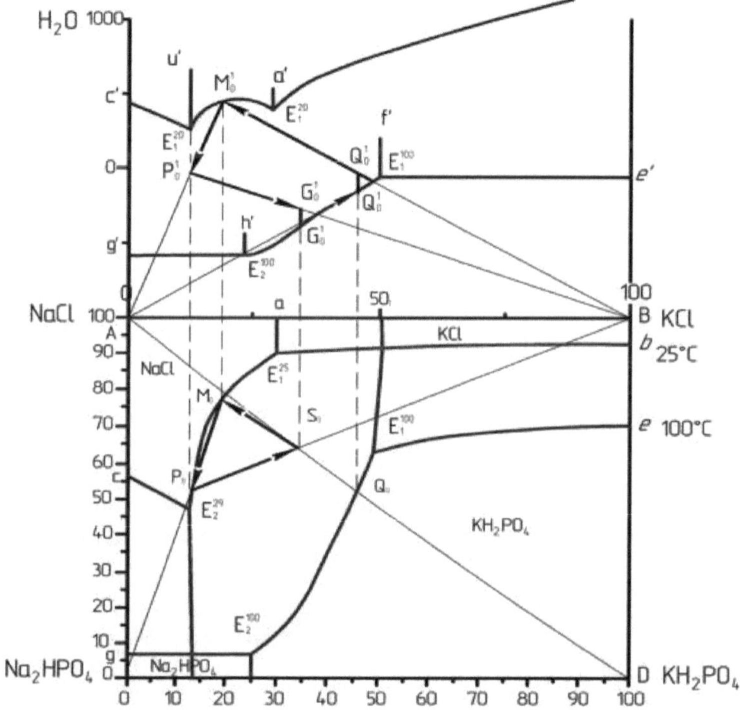

Figure 4.5. Diagram of K^+ , Na^+ // Cl^- , $H_2PO_4^-$ - H_2O as applied to the process of obtaining potassium dihydraphosphate

Table 4.1 summarizes the characteristics of suspensions in the production of potassium dihydrophosphate from a solution of sodium dihydrophosphate and flotation potassium chloride.

Table 4.1.

Characteristics of slurry in obtaining potassium dihydrophosphate from a solution of sodium dihydrophosphate and potassium chloride

№ Var.	Qo				Mo				Po	So	
	Suzp., d.	G.F.	T.F.	G:T.	Suzp., d.	G.F.	T.F.	G:T.	Suzp., d.	Suzp., d.	G.F.
1	189,35	170,73	18,62	9,17:1	183,33	142,49	40,84	3,49:1	323,68	300,21	189,35

2	189,74	169,93	19,71	8,62:1	180,76	137,12	43,64	3,15:1	330,38	305,35	189,74
3	191,31	170,84	20,47	8,35:1	181,67	136,71	44,96	3,04:1	336,01	310,20	191,31
4	190,63	170,15	20,48	8,31:1	184,58	139,62	44,95	3,11:1	338,93	313,11	190,63
5	189,35	170,74	18,61	9,17:1	183,96	143,09	40,87	3,50:1	324,28	300,81	189,35
6	182,77	167,26	15,51	10,79:1	196,09	162,03	34,06	4,76:1	313,0	293,45	182,77

Table 4.2 shows the results of experiments in obtaining potassium dihydrophosphate by six variants of the cyclic method.

Table 4.2.

Chemical composition of precipitates and mother liquors formed by at conversion of NaH_2PO_4 solution with KCl

№	Precipitation composition KH_2PO_4, %				Precipitation composition NaCl, %				Composition of mother liquor (M_o), %			
	P_2O_5	TO_2 ABOUT	NaO_2	Cl^-	P_2O_5	TO_2 ABOUT	NaO_2	Cl^-	P_2O_5	TO_2 ABOUT	NaO_2	Cl^-
1	42,05	29,15	0,89	0,99	0,64	0,95	52,24	43,71	4,18	6,04	10,88	6,13
2	43,76	28,87	0,12	0,81	0,73	0,14	52,75	44,42	4,93	10,44	11,41	6,85
3	48,85	33,02	1,66	0,62	1,99	1,70	60,23	56,83	6,42	4,41	12,69	7,99
4	43,95	28,30	8,99	3,49	3,90	3,04	52,74	44,61	5,77	5,53	13,08	8,70
5	39,52	25,29	10,62	4,54	6,35	3,11	57,82	42,26	11,15	4,08	14,22	7,60
6	43,72	22,40	8,66	3,41	6,93	2,71	41,87	40,34	17,17	3,10	11,59	7,20

After several cycles, potassium dihydrophosphate was isolated containing 48.85% P_2O_5, 33.02% K_2O, 1.66% Na_2O and 0.6% Cl^-. NaCl precipitate obtained by evaporating water from the solution contains

1.99% P_2O_5 , 1.70% K_2O, 60.23% Na_2O and 56.83% Cl^- . The mother liquor, which contains 6.42% P_2O_5 , 4.41% K_2O, 12.69% Na_2O is returned to the cycle. Next, the mono-potassium phosphate precipitate was washed with saturated mono-potassium phosphate solution and dried at 100-110°C.

The conducted analyses and calculations on solubility diagrams in the four-component system K^+ , Na^+ //Cl^- , $H_2PO_4^-$ -H_2O and the obtained experimental data showed the principal possibility of obtaining potassium dihydrophosphate by conversion method from solutions of monosodium phosphate obtained on the basis of EPC from CK phosphorites and flotation potassium chloride.

§ 4.3. Influence of technological parameters on the crystallization process of monokaliy phosphate

The influence of technological parameters such as solution concentration, temperature and cooling rate of the solution on the process of crystallization of mono-potassium phosphate from conversion solution obtained by conversion method from solutions of monosodium phosphate obtained on the basis of EPC from phosphorites CK and flotation potassium chloride was studied. The obtained results are summarized in Table 4.3.

A slow cooling rate favors the precipitation of fewer impurities. With decreasing cooling rate the amount of impurities Na_2O and Cl^- in the product decreases approximately twice. Increasing the solution concentration above 30 % leads to an increase in the impurity content in the finished product.

Table 4.3.

Degree of transition of solution components into product and chemical composition of mother liquor after potassium dihydrophosphate crystallization

Tempe-The nature of it, °C	Speed cooling, °C/hour	Degree of transition of components into the product, %				Chemical composition of mother liquor, wt. %			
		TO_2 ABOUT	PO_{25}	NaO_2	Cl^-	TO_2 ABOUT	PO_{25}	NaO_2	Cl^-

At a mono-potassium phosphate concentration of 27%									
31,0	10,0	68,48	69,72	4,28	2,59	3,98	5,72	11,42	7,20
	7,0	68,45	68,68	4,26	2,57	3,97	5,70	11,40	7,19
	5,0	68,43	68,66	4,25	2,56	3,97	5,70	11,40	7,18
25,0	10,0	79,62	81,06	4,60	2,78	3,93	5,72	10,15	7,11
	7,0	79,58	81,01	4,57	2,75	3,92	5,72	10,17	7,12
	5,0	79,56	80,99	4,55	2,74	3,92	5,72	10,18	7,13
At a monokaliy phosphate concentration of 24%									
31,0	10,0	61,14	62,25	3,82	2,31	4,47	6,50	12,83	8,08
	7,0	61,12	61,23	3,81	2,29	4,46	6,49	12,82	8,07
	5,0	61,11	61,21	3,79	2,28	4,46	6,49	12,82	8,07
25,0	10,0	71,09	72,38	4,11	2,48	4,41	6,42	12,69	7,99
	7,0	71,06	72,34	4,07	2,46	4,41	6,42	12,71	8,01
	5,0	71,04	72,32	4,05	2,44	4,41	6,42	12,72	8,02

The mother liquor after crystallization of potassium dihydrophosphate is returned for reuse in the initial stage on the conversion process of potassium chloride with sodium dihydrophosphate.

Thus, the optimum conditions for obtaining potassium dihydrophosphate by conversion of monosodium phosphate solution by flotation potassium chloride with preliminary release of sodium chloride are temperature - not more than 25.0 ° C, solution concentration - not less than 24-27%, solution cooling rate - 10.0 ° C/hour. The chemical composition of the obtained product meets the requirements of GOST.

§ 4.4 Investigations of monokaliy phosphate washing from impurities and chlorine

Table 4.4 shows the chemical and salt compositions of potassium dihydrophosphate obtained by single and double washing with saturated solutions of monosodium and monopotassium phosphates.

Table 4.4.

Chemical composition of potassium dihydrophosphate after washing

№	Ratio -nation KN_2 RO_4 :Prom	Ionic composition, %				Salt composition, %				Yield, %
		H_2 RO_4^-	K^+	Na^+	Cl^-	KN_2 RO_4	NaN_2 RO_4	KCL	NaCl	
NaH solution₂ RO₄										
1	1:1 NaH2PO4	70,92	27,08	1,30	0,70	94,42	4,43	-	1,15	95,79
2	1:1 NaH2PO4	70,82	27,62	0,92	0,64	96,33	2,62	-	1,05	95,46
3	1:1 NaH2PO4	70,82	27,62	0,92	0,64	96,33	2,62	-	1,05	95,46
1	1:2 NaH2PO4	70,87	26,97	1,40	0,76	94,07	4,67	-	1,26	87,30
2	1:2 NaH2PO4	70,78	27,52	1,00	0,70	95,97	2,88	-	1,15	87,00
3	1:2 NaH2PO4	70,78	27,52	1,00	0,70	95,97	2,88	-	1,15	87,00
KN solution₂ RO₄										
1	1:1KN2PO4	70,47	28,08	0,66	0,79	97,91	0,79	-	1,30	100,0
2	1:1KN2PO4	70,43	28,32	0,49	0,76	98,75	-	-	1,25	100,0
3	1:1KN2PO4	70,43	28,32	0,49	0,76	98,75	-	-	1,25	100,0
1	1:2KN2PO4	70,41	28,14	0,63	0,82	98,13	0,52	-	1,35	98,00
2	1:2KN2PO4	70,39	28,31	0,51	0,79	98,70	-	-	1,30	97,88
3	1:2KN2PO4	70,39	28,31	0,51	0,79	98,70	-	-	1,30	97,88

When KH_2 PO_4 crystals were washed with saturated NaH_2 PO_4 solution, a decrease in potassium phosphate yield for samples 1-3 was observed up to 95.46% at a ratio of KH_2 PO_4 :p-p = 1:1 and up to 87.00% at a ratio of 1:2, indicating dissolution of KH_2 PO_4 in saturated NaH_2 PO_4 solution and the greater the lower the ratio.

Washing with saturated solutions of KH_2 PO_4 of samples 1-3 practically does not lead to decrease of yield of KH_2 PO_4 at the ratio 1:1

and insignificantly decreases to 97,88-98,00% at the ratio 1:2. It follows from the obtained data that washing is preferably carried out with saturated solution of KH_2PO_4 at a ratio of 1:1.

§ 4.5. Rheological properties of initial suspension and stock solutions

For the experiments, the rheological properties of monosodium phosphate solution were studied at temperature 20-80°C and pH = 3.8 - 4.5. The results of the studies are summarized in Table 4.5.

Table 4.5.

Density and viscosity of monosodium phosphate solutions obtained by neutralization of EPC from CK phosphate rock with soda ash

№	pH	Density, g/cm^3				Viscosity, mPa·s			
		20°C	40°C	60°C	80°C	20°C	40°C	60°C	80°C
Initial solution of sodium dihydrophosphate									
1	3,8	1,240	1,234	1,230	1,228	3,41	2,17	1,42	1,10
2	4,1	1,242	1,236	1,232	1,230	3,50	2,23	1,49	1,14
3	4,5	1,246	1,240	1,236	1,234	3,96	2,55	1,68	1,29
Evaporated sodium dihydrophosphate solution									
4	3,8	1,389	1,382	1,378	1,375	3,82	2,43	1,59	1,23
5	4,1	1,391	1,384	1,380	1,378	3,92	2,50	1,67	1,28
6	4,5	1,396	1,389	1,384	1,382	4,39	2,86	1,88	1,45

As the pH increases from 3.8 to 4.5, the densities of the stock and evaporated solutions increase and are 1.240-1.246 and 1.389-1.396 g/cm^3 at 20° C and 1.228-1.234 and 1.375-1.382 at 80° C, respectively.

With increasing pH, the viscosities of the solutions also increase and are 3.41-3.96 and 3.82-4.39 mPa·s at 20° C and decrease to 1.10-1.29 and 1.23-1.45 mPa·s at 80° C, respectively. The initial and evaporated solutions of monosodium phosphate have good rheological properties.

Table 4.6 shows the values of density and viscosity of stock solutions formed after separation of potassium dihydrophosphate according to three variants [158; P. 261-265].

Table 4.6.

Density and viscosity of stock solutions formed after filtration of potassium dihydrophosphate

No. of rev from Table 4.2	Density, g/cm³				Viscosity, mPa·s			
	20°C	40°C	60°C	80°C	20°C	40°C	60°C	80°C
1	1,285	1,277	1,27 1	1,26 6	2,71 8	1,86 7	1,36 4	1,20 5
2	1,295	1,287	1,28 1	1,27 7	2,78 7	1,97 3	1,46 2	1,28 4
3	1,291	1,283	1,27 7	1,27 3	2,78 6	1,97 2	1,46 1	1,28 2

In all variants, the density and viscosity of mother liquors decrease with increasing temperature. The study of rheological properties of mother liquors shows that they have satisfactory mobility and can be easily transported.

§ 4.6 Physicochemical and commercial properties of monokaliy phosphate

In order to establish the salt and chemical composition of potassium dihydrophosphate, X-ray diffraction, IR spectroscopic and scanning electron microscopic studies were carried out. X-ray diffractometer XRD-6100 (Shimadzu, Japan) was used for X-ray analysis. CuK_α -radiation (β-filter, Ni, tube current and voltage mode 30 mA, 30 kV) and constant detector rotation speed of 4 deg/min were used. A rotational camera with a rotation speed of 30 rpm was used when the sample was taken. The X-ray radiograms were decoded using the American Mineralogist crystal structure database and Mikheev's X-ray determinant of minerals.

The IR absorption spectrum of the investigated substance - bischofite was taken on SPECORD-751R in the frequency range of 400-4000 cm^{-1} . Samples were prepared by pressing tablets with KBr.

Fig. 4.6 shows the X-ray diffraction and Fig. 4.7 IR spectrum of potassium dihydrophosphate synthesized under laboratory conditions.

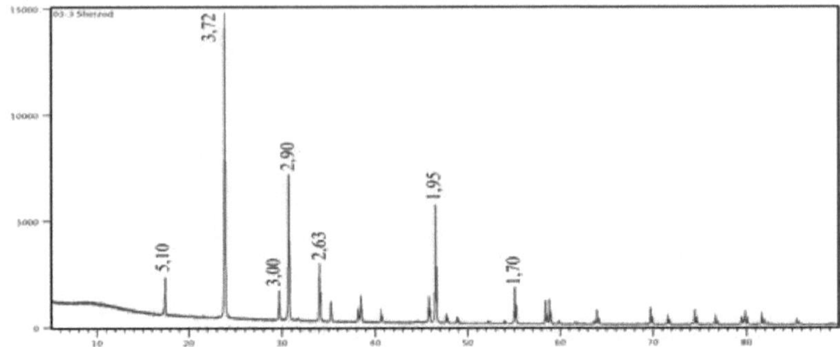

Figure 4.6. X-ray diagram of potassium dihydrophosphate

The X-ray radiograph of potassium dihydrophosphate shows diffraction maxima of 5.1; 3.72; 3.00; 2.90; 2.63 Å, which belong to KH_2 PO_4 , and also, 1.70 Å NaCl.

Figure 4.7. Infrared spectrum of potassium dihydrophosphate

On the IR - spectrum (Fig. 4.7) there are frequencies of vibrations characterizing vibrations related to RO_4 429,11-1065,79 cm^{-1} and crystalline water - 1277,42-2325,55 cm^{-1} . The data of IR spectroscopy of potassium dihydrophosphate confirm the data of chemical and X-ray studies.

Table 4.7 and Figure 4.8 show the main components of the obtained potassium dihydrophosphate based on sodium dihydrophosphate and flotation potassium chloride with conversion method. Scanning electron-microscopic analysis of potassium dihydrophosphate shows the following content of composition elements: O-43.31%, Na-0.78%, P-24.96%, Cl-0.48%, K-30.47%, which corresponds to their content in potassium dihydrophosphate.

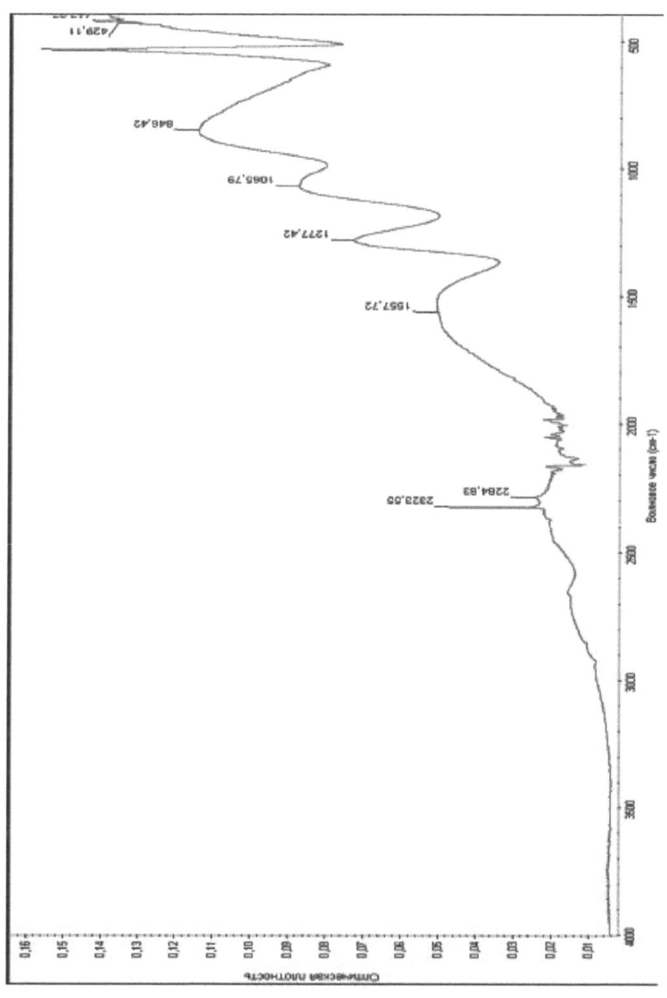

Talitsa 4.7

Results of elemental chemical analysis of monokaliy phosphate

Element	Weight. %	Sigma, Wt. %
O	43.31	0.23
Na	0.78	0.05

P	24.96	0.14
Cl	0.48	0.04
K	30.47	0.16
Summation:	100.00	

Электронное изображение 3

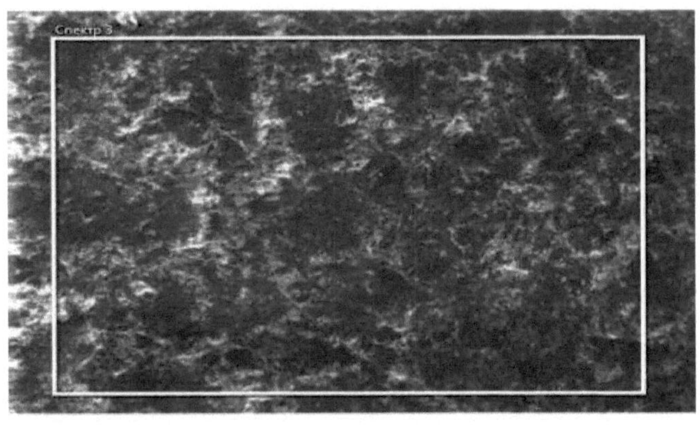

250µm

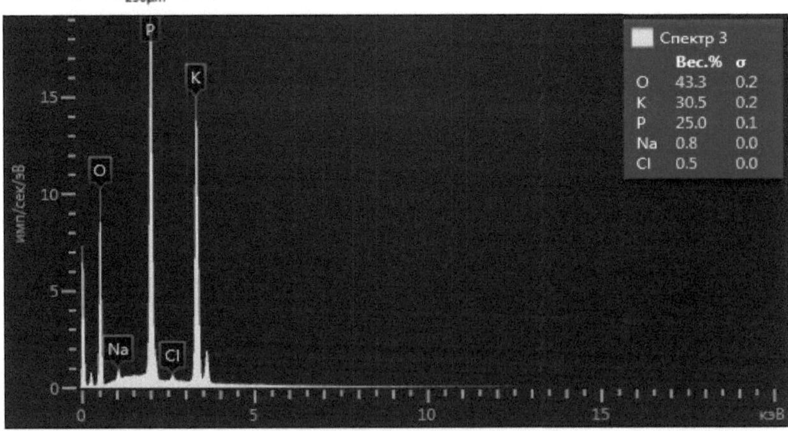

Figure 4.8. Scanning microscopic analysis of potassium dihydrophosphate

Thus, the influence of crystallization time and temperature on elemental, basic chemical and salt composition of potassium dihydrophosphate was studied. The obtained results were confirmed by X-ray phase and IR spectroscopic methods of analysis. It is shown that potassium dihydrophosphate obtained under optimal conditions after its washing with saturated potassium dihydrophosphate solution at the ratio KH_2 PO_4 :p-p = 1:1. The elemental composition of potassium dihydrophosphate was established by electron microscopy.

The results indicate the possibility of synthesizing potassium dihydrophosphate with the required performance properties through a solution of sodium dihydrophosphate and flotation potassium chloride.

§ 4.7 Technological scheme and material balance of monokaliy phosphate production

Technological scheme and material balance of production of chlorine-free mono-potassium phosphate of feed purity on the basis of EPCs obtained from phosphate rock are presented in Figures 4.9 and 4.10.

The reactor (item 1) is fed simultaneously with monosodium phosphate solution, fine crystalline potassium chloride at stirring and temperature 95-100°C to evaporate moisture and concentrate the mixture.

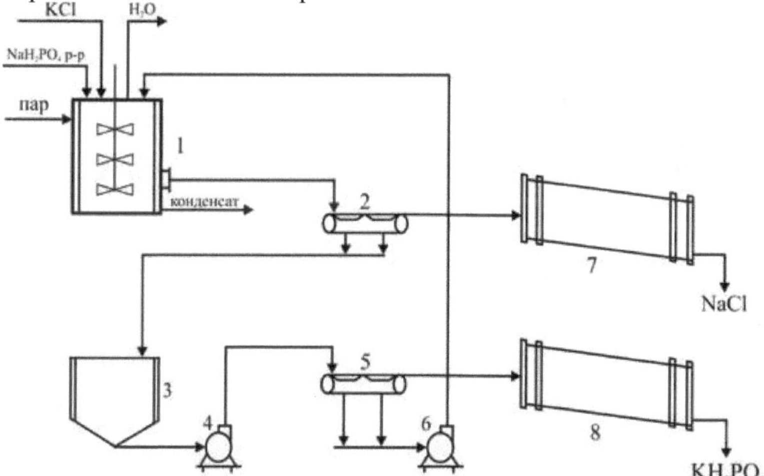

Fig. 4.9. Technological scheme for obtaining monokaliy phosphate of feed purity on the basis of extractive phosphoric acid

obtained from phosphorites of Central Kyzylkum by conversion method: 1 - reactor with jacket; 2, 5 - filters; 3 - crystallizer; 4, 6 - pumps; 7, 8 - dryers.

The reactor also receives the mother liquor after separation of mono-potassium phosphate. The conversion and evaporation processes take place within 2 hours, after which the mixture is fed for hot filtration to a vacuum filter (pos.2) to separate precipitated crystals of sodium chloride, which is sent for drying in a drying drum (pos.7). The filtrate from the filter (pos.2) goes to the crystallizer (pos.3) and then by pump (pos.4) to the filter (pos.5) for separation of crystals of monokaliy phosphate, which then goes to the drying drum dryer (pos.8) mother liquor is used as a circulation and is fed to the reactor (pos.1). Material balance of obtaining monokaliy phosphate on the basis of EPC obtained from phosphate rock CK is presented in Figure 4.10.

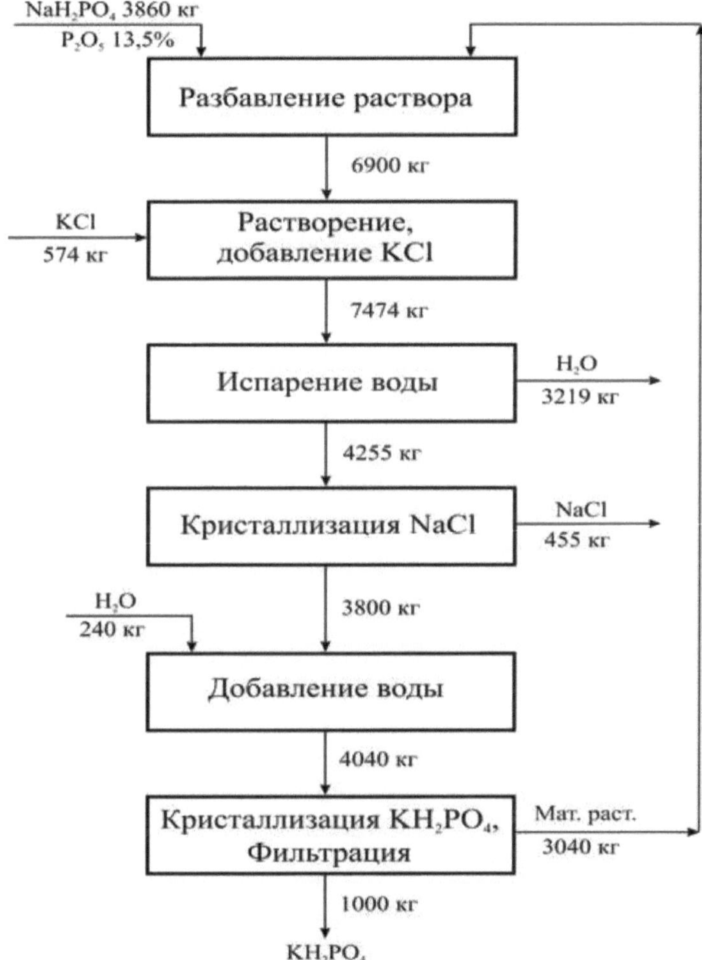

Fig. 4.10. Scheme of material flows and material balance of cyclic method of monokaliy phosphate production by conversion method

Due to the fact that the main task of the study was to obtain mono-potassium phosphate by conversion method, the material balance is made for the conversion stage.

To obtain 1000 kg of chlorine-free monocalcium phosphate feed purity should be in 3860 kg of purified solution of monosodium phosphate, containing 13.5% P_2O_5 enter 3040 kg of mother liquor after separation of monocalcium phosphate, and then 574 kg of potassium chloride and evaporate 3219 kg of water and filter the precipitate precipitated sodium chloride in the amount of 455 kg. Cool the mother liquor to 20-30°C and filter off the precipitate of monocalcium phosphate, wash with saturated solution of monocalcium phosphate and dry.

§ 4.8 Approval of the technology for monokaliy phosphate production

Conversion method of monokaliy phosphate production is a new product for the production facilities of the chemical industry of the republic, which requires different equipment from that available at the enterprises. Therefore, tests of the developed technology were carried out on a model plant in the joint venture JSC Elektrokimyozavod.

The technological process of monokaliy phosphate production consists of the following stages:

- Conversion of monosodium phosphate solution by potassium chloride;
- packing of the conversion solution;
- separation of precipitated sodium chloride;
- cooling and crystallization of mono-potassium phosphate;
- separation of the monokalium phosphate crystals;
- washing and drying of mono-potassium phosphate.

The technology was tested on a model plant simulating the production conditions and a pilot batch of monokaliy phosphate was produced.

Technological process. Monosodium phosphate solution is fed to the stage of conversion of monosodium phosphate with potassium chloride. The process of conversion and evaporation of the solution proceeds at a temperature of 100° C until precipitation of sodium chloride in the solution. The pulp is then fed for filtration. The solid phase - sodium chloride is sent for washing and drying to obtain a commercial product - sodium chloride, and the liquid phase in the crystallizer for cooling to a temperature of 20-30 ° C and crystallization of monokaliy phosphate. Precipitated crystals of monokaliy phosphate are separated from the mother liquor, washed and dried. The mother liquor is returned to the stage of monosodium phosphate conversion with potassium chloride.

As a result of technology testing, a pilot batch of monokaliy phosphate in the amount of 50 kg was produced and submitted for testing. Data on the analysis of chemical composition and marketable properties of the product are given in Table 4.8.

Table 4.8.

Chemical and physicochemical parameters and comparative characterization of obtained monokaliy phosphate by conversion method

№ n/a	Name of indicators	Indicators	
		according to specifications	In fact
1	Appearance	White powder colors	White powder colors
2	Mass fraction of total phosphates, (P_2O_5), %, not less than	50,0	50,0
3	Mass fraction of potassium, (K_2O), %, not less than	33,0	33,0
4	Mass fraction of moisture, %, max.	0,5	0,35
5	Hydrogen index of 1% solution, units. pH	4,0-4,5	4,3
6	Mass fraction of residue insoluble in water, %, max.	0,1	0,01

The mono-potassium phosphate obtained during the tests fully complied with the requirements. TU 2186-021-32496445-00 (RF).

Thus, the conducted tests showed the possibility of obtaining mono-potassium phosphate from monosodium phosphate solutions on the basis of purified EPC from CK phosphate rock by conversion with potassium chloride.

The tests carried out at the pilot plant show adequate reproducibility of the results of laboratory studies regarding the technological parameters of the process realization and the quality level of the obtained products.

§ 4.9. Preliminary technical and economic calculations of monokaliy phosphate production efficiency

In order to determine the feasibility of monokaliy phosphate production, preliminary technical and economic calculations were carried out.

Production of mono-potassium phosphate includes purification of EPC from fluorine and sulfate, neutralization of acid with soda ash to pH 4.0-4.5, separation of insoluble residues and obtaining purified monosodium phosphate solutions, conversion of flotation potassium chloride with monosodium phosphate solution, packing and separation of sodium chloride, cooling and crystallization of potassium dihydrophosphate, separation, washing and drying of potassium dihydrophosphate crystals, return of stock solutions to the initial stage of conversion.

The cost to produce mono-potassium phosphate consists of the cost of EPC, soda ash, liquid glass, MOFK, purified monosodium phosphate solution and flotation potassium chloride.

Prices for raw materials and energy resources are accepted: 1 ton of EPC with 17% P_2O_5 - 239480 soums, 1 ton of MOFC - 490539 soums (according to JSC "Ammophos-Maksam"), 1 ton of soda ash - 2076,824 soums, 1 ton of liquid glass 50% solution - 1500 thousand soums. soum, which is necessary to produce purified monosodium phosphate solution and 1 ton of flotation potassium chloride produced by JSC "Dehkanabad Potash Plant" - 1,635.090 thousand soums.

Prices for energy resources are taken 411 UZS per 1 kWh of electricity, steam 148596 UZS per 1 Gcal, recycled water 434477 per 1000 m^3 , compressed air 92790 UZS per 1000 m .[3]

To produce 1 ton of mono-potassium phosphate, 3.860 tons of purified monosodium phosphate solutions containing 13.5% P_2O_5 . Costs for 1 ton of purified monosodium phosphate solution with 13.5% P content$_2$ O_5 is 345,450 thousand soums.

The total cost to produce a purified monosodium phosphate solution is:

3,860 x 345,450 = 1,325,717 thousand soums.

Another 0.574 tons of potassium chloride is needed at a cost of 0.574 tons of potassium chloride to produce the mono-potassium phosphate conversion solution:

0,574 x 1635,090 = 938,542 thousand soums.

Total raw material costs will be:

1 325,717 + 938,542 = 2 264,259 thousand soums.

To produce 1 ton of monokaliy phosphate, 250 kW of electricity, 16.03 GJ or 3.829 Gcal of steam, 25 m^3 of compressed air and 66.67 m^3 of natural gas are consumed. Energy costs of monokaliy phosphate production will be as follows:

3.040 x 0.435 = 1.323 thousand soum cost of recycled water.

3,829 x 148,6 = 568,989 thousand soums cost of steam.

25 x 0.0928 = 2.32 thousand soum cost of compressed air.

66,67 x 380 = 25,335 thousand soums cost of natural gas.

250 x 0.411 = 102.75 thousand soums of electricity cost.

Total energy consumption for the production of 1 ton of monokaliy phosphate will be as follows:

1,323 + 568,989 + 2,32 + 25,346 + 102,75 = 700,713 thousand soums.

Expenses for raw materials and energy resources will amount to:

2264,259 + 700,713 = 2964,972 thousand soums.

Technological costs for monokaliy phosphate production are assumed to be 30% of the cost of raw materials and energy resources:

2964,972 x 0,30 =889,492 thousand soums.

Technological cost of monokaliy phosphate will be as follows:

2964,972 + 889,492 = 3854,464 thousand soums.

Period expenses and sales of monokaliy phosphate are assumed to be 10% of the technological cost:

3854,464 x 0,1 = 385,446 thousand soums.

The full factory cost of production will be:

3854,464 + 385,446 = 4239,910 thousand soums.

Wholesale price of mono potassium phosphate 1600 USD per 1 ton or 1600 x 10855 = 17,368.00 thousand soums without transportation and customs costs, where 10855 soums dollar exchange rate (20.12.2021) US dollars per 1 ton or 1600 x 10855 = 17,368.00 thousand soums without transportation and customs costs, where 10855 soums dollar exchange rate in soums (20.12.2021).

Savings from each ton of mono-potassium phosphate will amount to 17,368.00 - 4,239.910 = 13,128.09 thousand soums, compared to imported: 17,368.00 - 4,239.910 = 13,128.09 thousand soums.

In addition, the production of 1 ton of monokaliy phosphate will produce 0.5 tons of sodium chloride, the cost of which will be:

0.5 x 500.0 = 250.0 thousand soums.

With the production of 10 thousand tons of crystalline monokaliy phosphate and 5 thousand tons of sodium chloride, the total savings will amount to 15,628.09 million soums.

Thus, the conducted technical and economic calculations indicate good profitability of monokaliy phosphate production by conversion of sodium dihydrophosphate solution obtained on the basis of EPC from phosphate rock with flotation potassium chloride produced by JSC "Dehkanabad Potash Plant", high efficiency of production, allowing to obtain import-substituting chemical products, create additional jobs.

§ Conclusions to Chapter 4

For physicochemical substantiation of the possibility of obtaining monokaliy phosphate, theoretical and experimental studies were carried out, which made it possible to develop the technology of chlorine-free phosphorus-potassium fertilizers by conversion of monosodium phosphate with potassium chloride.

The crystallization field of $KN_2 PO_4$ on the solubility diagram of the system K^+ , Na^+ // Cl^- , $H_2 PO_4^-$ - $H_2 O$ at 25°C occupies a larger part of the square area than at 100°C. When considering the isothermal solubility diagram, it can be seen that as the temperature increases, the crystallization fields of $KN_2 PO_4$ decrease sharply, hence their solubility increases significantly. The saturation region of $CH_2 PO_4$ at 100°C is several times smaller than at 25°C.

The crystallization field of NaCl at 100°C is significantly increased, compared to the crystallization field at 25°C, due to a decrease in the field of $CH_2 PO_4$. Consequently, the solubility of $KN_2 PO_4$ is increased several times and such a solution is capable of precipitating this salt on cooling.

The CH $system_2$ PO_4 -NaH_2 PO PO_4 -NaCl- KCl-H_2 O has been used to analyze the phase composition of industrial processes.

On the basis of analysis of phase equilibria using the solubility diagram of four-component water-salt system K^+ , Na^+ // Cl^- , $H_2 PO_4^-$ - H_2 O at 25°C and 100°C the data on the possibility of isolation of potassium dihydrophosphate and sodium chloride by their crystallization from saturated solutions were obtained.

The influence of technological parameters such as solution concentration, temperature and cooling rate of the solution on the process of crystallization of mono-potassium phosphate from conversion solution obtained by conversion method from solutions of monosodium phosphate obtained on the basis of EPC from CK phosphate and flotation potassium chloride has been studied.

A slow cooling rate favors the precipitation of fewer impurities. With decreasing cooling rate the amount of impurities Na_2 O and Cl^- in the product decreases approximately twice. Increasing the solution concentration above 30 % leads to an increase in the impurity content in the finished product.

Optimal conditions for obtaining potassium dihydrophosphate by conversion of monosodium phosphate solution by flotation potassium chloride with preliminary release of sodium chloride are temperature - not more than 25.0 ° C, solution concentration - not less than 24-27%, solution cooling rate - 10.0 ° C/hour. The chemical composition of the obtained product meets the requirements of GOST.

Rheological properties of conversion solutions at temperatures of 20-80°C depending on pH were studied. With increasing pH from 3.8 to 4.5, the densities of the initial and evaporated solutions increase and are 1.240-1.246 and 1.389-1.396 g/cm^3 at 20° C and 1.228-1.234 and 1.375-1.382 at 80° C, respectively.

With increasing pH, the viscosities of the solutions also increase and are 3.41-3.96 and 3.82-4.39 mPa·s at 20° C and decrease to 1.10-1.29 and 1.23-1.45 mPa·s at 80° C, respectively. The initial and evaporated solutions of monosodium phosphate have good rheological properties.

On the basis of the performed researches the basic technological scheme of monokaliy phosphate and sodium chloride production is offered. The optimum parameters of technological process are established and material balance and norms of technological regime are compiled.

The tests carried out at the pilot plant show adequate reproducibility of the results of laboratory studies regarding the technological parameters of the process realization and the quality level of the obtained products.

LIST OF REFERENCES

1. Decree of the President of the Republic of Uzbekistan No. UP-4947 dated February 7, 2017 "Strategy of actions on five priority directions of the Republic of Uzbekistan in 2017-2021".

2. Decree of the President of the Republic of Uzbekistan No. PP-4265 dated April 3, 2019 "On measures for further reforming and increasing the investment attractiveness of the chemical industry".

3. Decree of the President of the Republic of Uzbekistan No. PP-4937 dated December 28, 2020 "On measures for implementation of the investment program of the Republic of Uzbekistan for 2021-2023".

4. Resolution of the President of the Republic of Uzbekistan No. PP-4005 dated November 6, 2018 "On additional measures for further development of the fish farming industry".

5. Van Weser J. Phosphorus and its compounds. - Moscow: 1962, Vol. 1. - 687 p.

6. Melikulova G.E., Mirzakulov H.Ch., Usmanov I.I., Isakov A.F. Study of the process of obtaining feed dicalcium phosphate from phosphorites of the Central Kyzylkum // Universum: technical sciences: electronic scientific journal. 2018. № 6(51). URL: https://7universum.com/ru/tech/archive/item/6037.

7. Beglov B.M., Ibragimov G.I., Sadykov B.B. Non-traditional methods of processing phosphate raw materials into mineral fertilizers. // Chemical Industry. -2005. T. 82. -№ 9. -C. 453-468.

8. Phosphorus - "element of life", its increasing role for mankind // Phosphates at the turn of XXI century. - Moscow, Almaty, Zhanatas. 2006. 201 c.

9. Reference book on feeding of farm animals. - Moscow: Rosselmash. 1983. - C. 69-78.

10. Belokon L.M., Bogdanova N.S., Mikhaleva T.K., Dokholova L.D. Tendencies of development of production of fodder dicalcium phosphate and complex mineral additives for animal husbandry on its basis. Mineral Fertilizers Industry. Series: Mineral fertilizers and sulfuric acid. Review information. NIITEKHIM. 1987. - 38 c.

11. Degtyarev V. Efficiency of monocalcium phosphate in animal feeding. Dairy and beef cattle breeding. 2003. - № 2. - C. 7-10.

12. Arifjanova K.S., Mirzakulov H.Ch., Melikulova G.E., Huzhamkulov S.Z., Usmanov I.I. Study of the process of desulfurization of extractive phosphoric acid from phosphorites of the Central Kyzylkum

// Journal of Chemical Industry - St. Petersburg, RF. 2017. № 5. - C. 18-25.

13. Gorlov I.F., Randelin D.A., Struk A.N., Struk V.N., Struk M.V., Struk N.V. Innovative technologies for the development and use of new feed and biologically active additives in the production of meat of farm animals and poultry. - Volgograd: FGBOU VPO Volgograd GAU, RF. 2012. - 236 c.

14. Vinogradov V.N., Duborezov V.M., Kirilov M.P. Feeding and fodder production in dairy cattle breeding // Zh. Achievements of science and technology in PAC. - 2009. - № 8. - C. 33-35.

15. Ryabov N.I., Levakhin V.I., Zelepukhin A.G., Korolev L., Popov V.V. et al. Increase of efficiency of fattening of young cattle. - MOSCOW: RF. 2005. - 109 c.

16. Pozin M.E.. Technology of mineral salts (fertilizers, pesticides, industrial salts, oxides, acids), 2.1, ed. 4th, corrected. - L.: Khimiya Publishing House, 1974. - 1556 c.

17. Technology of phosphorus and complex fertilizers, edited by S.D. Evenchik, A.A. Brodsky. - Moscow: Khimiya, 1987. - 464 c.

18. Litusova N.M. Technology of obtaining fodder calcium phosphates in granulated form on the basis of chalk and extraction phosphoric acid. Dissertation. ... candidate of technical sciences. Moscow, 2004. - 97 c.

19. Kemira Animal Nutrition's forecast for the feed phosphate market through 2015. - 13 c.

20. Sleen J. Phosphorus availability m 21st century/ Menagement of a nonrenewable resourse // Phosphorus and potassium. 1998. JVB 217. - pp. 25-31.

21. Production of solids and other inorganic substances // Information and technical reference book on the best available technologies.

22. Bushuyev N.N. Physico-chemical bases of influence of impurities of phosphate raw materials in technology of phosphorus-containing mineral fertilizers and pure substances. Diss. ... D. Sc. 2000. 412 c.

23. Hijran Z. Toama World phosphate industry // Iraqi Bulletin of Geology and Mining Special Issue, No.7, 2017, pp. 5 - 23. 5 - 23.

24. Umarov Sh.I., Melikulova G.E., Usmanov I.I., Mirzakulov H.Ch. Study of the process of processing of phosphoric acid solutions of phosphate concentrate enrichment of Central Kyzylkum // Universum:

Technical Sciences: electronic scientific journal. 2018. no. 6(51). URL: http://7universum.com/ru/tech/archive/item/6088.

25. Karmyshov V.F., Sobolev B.P., Nosov V.N. Production and application of fodder phosphate - M.: Khimiya, - 1987. - 272 c.

26. Albatyrov I. Place and role of the agrarian sector in the world economy // International Agricultural Journal. MOSCOW: RF. - 2005, №2. - C. 28-30.

27. Beglov B.M., Namazov Sh.S. Phosphorites of Central Kyzylkum and their processing. - Tashkent, 2013. - 416 c.

28. Gordeev A.V. Food security - the problem of the XXI century. // Collection of reports of the international conference. "Food security of Russia". March 12-14, 2002, Moscow. - Moscow: FGNU "Rosinformagroteh". 2002. - №1. - C. 94-96.

29. Mirzakulov H.Ch. Physico-chemical bases and technology of processing of phosphorites of Central Kyzylkum. Tashkent, Navruz Publishing House. 2019, 416 c.

30. Turaev Z. Development of technology of single phosphorus and complex fertilizers with microelements on the basis of phosphorites of Central Kyzylkum. Dissertation (DSc). Tashkent, 2021, 197 p.

31. Turaev Z., Shamshidinov I.Sh., Usmanov I.I. Technology of single phosphate and complex fertilizers with trace elements. - Namangan, 2020, 160 p.

32. Patent No. IAP 06411 (UZ). Method of obtaining liquid bioorganic-mineral fertilizer. I.T.Shamshidinov, T.T.Shamshidinov, I.I.Usmanov, G.K.Kadirova. - Publ.2021, Bulletin No.2.

33. Patent No. IAP 06263 (UZ). Method of preparation of ammonium phosphates. I.T.Shamshidinov, I.I.Usmanov, H.Ch.Mirzakulov, B.A.Mamurov, G.K.Kodirova - Published. 2020. Bul. no. 8.

34. Aguilar K.K. Preparation of potassium dihydrophosphate by conversion method. Dissertation. ... candidate of technical sciences. St. Petersburg, 2004, 187 p.

35. Dormeshkin O.B., Vorobyev N.I. Production of chlorine-free water-soluble complex fertilizers. Minsk, 2006, 248 p. ISBN985-434-623-4.

36. Shatilov V.I. Waste-free technology of obtaining water-soluble chlorine-free NPK-fertilizers on the basis of potassium phosphate. Dissertation. Candidate of Technical Sciences. Minsk, 2004. - 265 c.

37. Volkov A.V. Mineral fertilizer market. IV quarter, 2015. National Research University. Higher School of Economics. RF development center. 2015. - 67 c.

38. World fertilizer trends and outlook to 2022 Food and agriculture organization of the united nations. Rome, 2019. 28 p. https://www.fao.org/3/ca6746en/ca6746en.pdf

39. Short-Term Fertilizer Outlook 2019 - 2020 Market Intelligence and Agriculture Services International Fertilizer Association (IFA) IFA Strategic Forum 18-20 November 2019 Versailles (France) https: 2019_ifa_strategic_forum_versailles_public_summary.pdf

40. Nedelciu C.E., Ragnarsdottir K.V., Schlyter P., Stjernquist I. Global phosphorus supply chain dynamics: Assessing regional impacts to 2050 Global Food Security 26 (2020). 10 p. 100426 www.elsevier.com/locate/gfs

41. Uriadov D.N. Efficiency of optimization of feeding rations for lactating cows of Kulunda intra-breed type of red steppe breed. Diss. ... c.s.n. Barnaul. RF. 2012. 171 c.

42. State Register Report. No. I-2015-7-2 "Experimental industrial development of technology of fodder potassium and calcium phosphates on the basis of extract phosphoric acid of Central Kyzylkum". - Tashkent, 2016, 97 p.

43. Samadiy M.A. Technology of obtaining potassium and sodium chlorides from low-grade Tyubegatan sylvinites and halite wastes. Diss. ... Doctor of Phil. (PhD) in technical sciences. Tashkent, 2017. 142 c.

44. Nabiev M.N., Osichkina R.G., Tukhtaev S.T. Potassium sulfate with microelements. Tashkent, Izdvo: "Fan", 1998. 164 c.

45. Ivanova S.I. Technology of potash fertilizer production. 2013 http://eeca-ru.ipni.net/article/EECARU-2166.

46. Andronov V.I., Brodsky A.A., Zabeleshinsky Y.A. et al. Thermal phosphoric acid, salts and fertilizers on its basis. Moscow. 2019, 205 c.

47. Jančaitienė K. Sustainable technology of potassium dihydrogen phosphate production and liquid waste recovery // Summary of Doctoral Dissertation Technological Sciences, Chemical Engineering (05T). Kaunas, Germany. 2017, 36 p.

48. Dormeshkin O.B., Vorobyev N.I., Shatalo V.I. Waste-free technological process of obtaining chlorine-free water-soluble complex fertilizer based on potassium phosphate. // Chemical Technology, Minsk, Belarus. 2014. - T.15. № 6. - C. 24-32.

49. Vorobyev N.I. Technology of phosphorus and complex fertilizers. - Minsk: BSTU, 2015. - 177 c. ISBN 978-985-530-432-7.

50. Volkova A.V. Market of mineral fertilizers // 2019. 52 c. Market%20of%20mineral%20fertilizers-2019.pdf

51. Kisilev V.G. Production of monocalcium phosphate from poor phosphate raw materials by recirculation scheme. Dissertation. ... Candidate of Technical Sciences Moscow, 2013. 165 c.

52. Klassen P.V., Zavertyaeva T.I., Adamov E.A., Milkov G.A., Razmakhnina G.S. (Russia, JSC "NIUIF", Moscow) Use of poor phosphate raw materials for phosphate fertilizers // Chemical Industry Today. 2003, № 12, - C. 4-8.

53. Patent No. 2411222 (Russian Federation). Method of preparation of dicalcium phosphate. Sharipov T.V., Mustafii A.G. Opubl.10.02. 2011. Bul. 4.

54. Patent No. 2373144 (Russian Federation). Method of fodder dicalcium phosphate production. Dmitrevsky B.A., Treuschenko N.N., Lavrova T.V. Published. 20.11.2009. Bulletin. No. 32.

55. Vandysheva A.A. Resource-saving technologies in the production of defluorinated feed phosphate // Young Scientist. - 2015. - № 23.1 (103.1). - C. 16-19.

56. Filenko I.A. Acid decomposition of natural phosphorites with obtaining various forms of complex fertilizers. Diss. ...k.t.n. Moscow, 2019. 141 c.

57. Melikulova G.E. Development of technology of fodder ammonium and calcium phosphates from phosphorites of Central Kyzylkum. Diss. ....d.f. (RhD) by t.n. Tashkent-2018. 107 c.

58. Olifson A.L., Makhov S.V. Research in the field of complex processing of phosphate raw materials. Comparative analysis of researchers' works for 1970-2012. M.: 2015. - 426 c.

59. Melikulova G.E., Erkaev A.U., Toirov Z.K., Mirisaeva D.A. Physicochemical properties of pulps formed during the production of feed pritzipitat conversion of ammonium monophosphate by calcium nitrate // Chemistry and Chemical Technology. -Toshkent, 2012. - № 4. - C. 2-5.

60. Patent No. IAP 05054 (UZ). C05B3/00. Method of preparation of feed precipitate. Kh.Ch.Mirzakulov, I.I.Usmanov, B.B.Sadykov, N.V.Volynskova, G.E.Melikulova, Sh.I.Umarov. Published. 2015. Bulletin. no. 7.

61. Alimov U.K., Namazov Sh.S., Reimov A.M., Kaymakova D.A. Flow-circulation method of production of double superphosphate on the

basis of phosphate rock from the Central Kyzylkum // Chemical Industry. - St. Petersburg, 2015. - т. 92. № 3. - С. 109-118.

62. Alimov U.K., Rasulov A.A., Namazov Sh.S., Kaymakova D.A. Use of mineralized mass of phosphate rock of Central Kyzylkum in the process of production of double superphosphate by cyclic method // Chemical Industry. - St. Petersburg, 2017. - т. 94. № 1. - С. 1-10.

63. Safranova T.V., Sadilov I.S., Chaikun K.V., Shatalova T.B., Filippov Ya.Yu. Synthesis of monetite from calcium hydroxylamate and monocalcium phosphate monohydrate under conditions of mechanical activation. Journal of Inorganic Chemistry, Moscow, RF. 2019, Vol. 64, No. 9, - P. 916-922.

64. Ryadchikov V.G. Fundamentals of nutrition and feeding of farm animals. - Krasnodar, RF. 2012. - 328 c.

65. Egorov IA, Andrianova EN, Grigorieva EN Use of defluorinated phosphate and monocalcium phosphate in mixed fodder for broiler chickens and laying hens. // Poultry farming. - №04, 2020. - С. 27-32.

66. Patent of the Russian Federation. № 2461517. Method for obtaining dicalcium phosphate Sharipov T.V., Mustafin A.G. Application 2010154807/05, 30.12.2010, Publ. 20.09.2012. Bulletin No. 26.

67. Sultanov B.E., Namazov S.S., Zakirov B.S. Study of chemical enrichment of phosphate flour of Central Kyzylkum // Chemical Industry (St. Petersburg). - 2013. - Т. 90, №2. - С. 79-86.

68. Sultonov B.S., Namazov Sh.S., Zakirov B.S. Investigation of nitric acid benefication of low grade phosphorites from Central Kyzylkum // Journal of Chemical Technology and Metallurgy, 50, 1, 2015, pp. 26-34.

69. Sultanov B.E., Namazov S.S., Zakirov B.S. Beglov B.M. Effect of the concentration of calcium nitrate solution on the degree of washing of phosphoconcentrates obtained during chemical enrichment of highly carbonized phosphorites of the Central Kyzylkum // Reports of the Academy of Sciences of the Republic of Uzbekistan. - 2013. №1, - С. 51-54.

70. Olifson A.L., Makhov S.V. Investigations in the field of complex processing of phosphate raw materials. Moscow, 2015. 426 c.

71. Rodin V.I., Litusova N.M., Mikhaleva T.K. et al. Investigation of the process of obtaining feed calcium phosphates using high-speed mixing methods. // Proceedings of NIUIF, 2004. - С. 185-186.

72. Almukhametov I.A. Development and industrial development of technology of calcium dimonophosphate dissertation of Candidate of Technical Sciences, Moscow, 2000, 159 p.

73. Rodin V.I., Litusova N.M., Levin B.V., Kazakov A.I. Obtaining granulated feed monocalcium phosphate with the use of turbolopast mixer-granulator. // Chemical Industry Today. No. 12, Moscow, 2004,

74. Patent No. 2256607 (RU). Method of preparation of monocalcium phosphate. B.V.Levin, N.M.Litusova, V.I.Rodin, A.M.Kershner et al. Published. 20.07.2005. Bulletin No. 20.

75. Patent No. 2136637 (RU). Method of obtaining feed calcium phosphate and a line for its production. // Atazhakhova S.P. Zaidav. 2006128525/13, 04.08.2006, Publ. 27.06.2008. Bulletin No. 18

76. Patent No. 2255042 (RU). Method of obtaining monocalcium phosphate. // Brodsky A.A., Rodin V.I., Levin B.V., Litusova N.M., Grishaev I.G., Grinevich V.A., Davydenko V.V. Appl. 2004124216/15, 10.08.2004, Publ. 27.06.2005. Bulletin No. 18

77. Vandysheva A.A., Gafurov M.A. Resource-saving technologies in the production of defluorinated feed phosphate. Journal "Young scientist". 2015, № 23.1(103.1), - C. 16-17.

78. Myrzakhmetova B.B., Besterikov U.B., Petropovlovsky I.A. Obtaining monocalcium phosphate from Karatau phosphorites. www.rusnauka.com/1 1_NPE_2012/Chimia/7_108158.doc.htm.

79. Kiselev V.G., Pochitalkina I.A., Petrolovivsky I.A. Obtaining monocalcium phosphate from low-grade phosphate raw materials. Journal "Uspekhi chemii i khimicheskoy tekhnologii". Vol. XXIV, 2010, No. 9, - P. 77-80.

80. Myrzakhmetov B.B., Besterekov U.B., Petropavlovskiy I.A. Obtaining double superphosphate from phosphorites Kokdzhon and Koksu by liquid-phase method. United Scientific Journal. - Moscow, 2012, No. 2. - C. 60-64.

81. Myrzakhmetov B.B., Besterekov U.B., Petropavlovsky I.A., Pochitalkina I.A., Kiselev B.G. Kinetic regularities of decomposition of low-grade phosphorites by liquid-phase method under conditions of mother liquor recycling. Chemical industry today. - Moscow, 2012. № 5, - C. 6-9.

82. Kiselev V.I. Production of monocalcium phosphate from poor phosphate raw materials by circulation scheme. Abstract dissertation ... candidate of technical sciences. - Moscow, 2013. 16 c.

83. Akhmetova S.O., Moldobekov Sh.M. Technology of processing of low-grade phosphorites Karatau on monocalcium phosphate. Science News of Kazakhstan. - Shymkent, 1999, № 2. - C. 11-15.

84. Moldabekov K.T., Zhantasov J.K., Zhanmoldaeva J.M., Balabekov O. Kinetics of decomposition of low-quality phosphorites phosphoric acid and obtaining double superphosphate by cyclic method. Modern knowledge-intensive technologies. 2013, № 11, - C. 107-112.

85. Alimov U.K., Rasulov A.A., Namazov Sh.S., Reimov A.M., Kaimanova D.K. Optimal mode of processing of phosphate rock from Central Kyzylkum with packed extraction phosphoric acid. Universum: Technical Sciences: electronic scientific journal. 2016. № 8 (29). URL: http://7universum.com/ru/tech/archive/item/3572.

86. Alimov U.K. Development of resource-saving technology of highly concentrated phosphorus-containing fertilizers on the basis of phosphorites of Central Kyzylkum, Dissertation. ... doctoral candidate in technical sciences. Tashkent, 2019. 229 c.

87. Kiselev V.G., Ryashko A.I., Pochitalkina I.A., Petropavlovskiy I.A. Features of acid processing of phosphate raw materials of Polpinskoye deposit. Journal "Uspekhi khimiya i khimicheskoy tekhnologii". Vol. XXУ, 2011, № 8 (24). - C. 65-69.

88. Petropavlovskiy I.A., Pochitalkina I.A., Kiselev V.G., Ahnazarova S.L., Myrzakhmetova B.B. Obtaining monocalcium phosphate from poor phosphate raw materials by liquid-phase circulation method. Journal of Chemical Technology. 2012, № 8, - C. 453-456.

89. Kochetkov N.V. Reference book on phosphorus-containing fertilizers. - Moscow: Khimiya. 1982. - 400 c.

90. Zaitsev I.D., Aseev G.G. Reference book on physico-chemical properties of binary and multicomponent solutions of inorganic substances. - Moscow: Khimiya. 1988. - 416 c.

91. Jančaitienė K. Sustainable technology of potassium dihydrogen phosphate production and liquid waste recovery // Summary of Doctoral Dissertation Technological Sciences, Chemical Engineering (05T). Kaunas, Germany. 2017, 36 p.

92. Sotiboldiev B.S., Khoshimkhanova M.A., Dehkanov Z.K., Aripov H.S. Technology for obtaining new complex phosphorus fertilizers // Universum: Chemistry and Biology: electronic scientific journal. 2020. № 6(72). URL: http://7universum.com/ru/nature/archive/item/9375.

93. Dormeshkin O.B., Vorobyev N.I., Shatalo V.I. Waste-free technological process of obtaining chlorine-free water-soluble complex fertilizer based on potassium phosphate. // Chemical Technology, Minsk, Belarus, 2014. - T.15. № 6. - C. 324-332.

94. Knunyants I.L. Chemical Encyclopedia. - Moscow: Bolshaya Ross. Encyclopedia. 1992. - T.III. - C. 366-367.

95. Khuzhamberdiev S.M., Arifjanova K.S., Mirzakulov H.Ch. Study of the process of obtaining solutions and sodium salts suitable for the production of polyphosphates // Journal of Chemistry and Chemical Technology. - Tashkent, 2018. - № 3. - C. 18-21.

96. Mazunin S.A., Chechulin V.L., Frolova S.A., Kistanova N.S. Technology of obtaining potassium dihydrophosphate in the system with salting // Chemical Industry. RF. 2010. №1. - C. 6-15.

97. Patent No. 2164494. Method of obtaining monokaliy phosphate / P.N. Novikov, K.N. Ovchinnikova, R.I. Umansky. - No. 2000109523/12; Appl. 18.04.2000; Published 27.03.2001 // Inventions. Useful models. 2001. № 8.

98. Kochetkov S.P., Smirnov N.N., Ilyin A.P. Concentration and purification of extraction phosphoric acid. Ivanovo, 2007, 304 p.

99. Galtzov A.V., Zhelyaletdinova R.A. Calculation of water vapor pressure over super-phosphoric acid // Mineral fertilizers and sulfuric acid. - M.: NIITEKHIM, 1982. -Vyp. 2. - C. 15-17.

100. Patent No. 2608017 of the Russian Federation. Method for obtaining soluble chlorine-free potassium fertilizers / Khamizov R.H., Vlasovkikh N.S., Egorov V.G., Moroshkina L.P., Petukhov M.A., Smirnov A.A., Khamizov S.H. // Applic. 14.07.2015. № 2015128269. Published 11.01.2017. Bulletin No. 2.

101. Mazunin S.A., Chechulin V.L., Frolova S.I., Kistanova N.S. Technology of Obtaining of Potassium Dihydro-phosphate in the System with Salting-Out // Russian Journal of Applied Chemistry. 2010. Vol. 82. No. 3 (March), pp. 553-561.

102. Mazunin S.A., Chechulin V.L.. Applied Aspects of Use of Amines for the Production of Inorganic Salts in Systems with Salting-out // Russian Journal of Applied Chemistry. 2010. Vol. 83. No. 9, pp. 1690-1697.

103. Mazunin S.A., Chechulin V.L. Desalination as a physical and chemical basis for low-waste methods of obtaining potassium and ammonium phosphates. Monograph / Perm, 2012. - 114 c.

104. Patent No. 2712689 Russian Federation. Method for obtaining high-purity potassium dihydrophosphate // Komendo I.Yu., Zharova A.A. Application: 2018142161, 29.11.2018 Opub. 30.01.2020 Bulletin No. 4.

105. Patent No. 2261222 (RU). Method of obtaining monokaliy phosphate / Bogach E.V., Levin B.V., Rakcheeva L.V., Ovchinnikova K.N., Kremenetskaya E.V. Appl. 10.08.2004, Opubl. 27.09.2005.

106. CN 102963874 A China Method for producing industrial-grade potassium dihydrogen phosphate (KH_2 PO_4) by wet method purified phosphoric acid and potassium sulfate. 2013.03.13

107. Patent No. 10538 BY. Method of obtaining monokaliy phosphate // Shablovsky V.O., Tuchkovskaya A.V., Pap O.G., Chernevich N.P., Baltrushevich A.G., Surtaev A.F., Starovoitov S.M. Zaidav. 20041103, 2004.11.29. Published. 2006.06.30.

108. Dormeshkin O.B., Novik D.M., Shatilo V.I. Universal technology for obtaining chlorine-free water-soluble complex fertilizers based on technical products // Vestnik PNIPU, Minsk, Belarus, 2018, № 4, - P. 163-173.

109. Vorobyev N.I., Dormeshkin O.B., Shatilo V.I. Obtaining chlorine-free water-soluble NPK fertilizers by conversion method // Vesci National Academy of Sciences of Belarus. Ser. chim. nauk. - 2004. - № 1. - C. 96-101.

110. Patent No. 2747639 Russian Federation. Method of obtaining monokaliy phosphate // Olifson A.L., Kapustinsky N.N. Application: 2020121330, 26.06.2020, Opub. 11.05.2021, Bul. No. 14

111. Patent No. 2103348 of the Russian Federation. Method of obtaining potassium ammonium phosphate solution // Stepchenko A.G., Sorokin G.V. Application: 94038480, 10.10.1994, Published. 27.01.1998.

112. Patent No. 2633569 Russian Federation. Continuous method of production of neutral granular phosphorus-potassium fertilizer // Aviv T., Cohen U.S., Orgil D., Aroh I. Application: 2014152889, 27.06.2013, Opub. 13.10.2017. Bulletin No. 29.

113. Arifjanova K.S., Khuzhamkulov S.Z., Normurodov B.A., Shamaev B.E., Mirzakulov H.Ch. Desulfurization of extraction phosphoric acid from phosphorites of Central Kyzylkum by unenriched phosphate raw material // Journal "Chemical Technology. Control and management" - Tashkent, 2016. № 4. - C. 27-33.

114. Arifjanova K.S., Mirzakulov H.Ch., Melikulova G.E., Huzhamkulov S.Z., Usmanov I.I. Study of the process of desulfurization of extractive phosphoric acid from phosphorites of the Central Kyzylkum

// Journal of Chemical Industry - St. Petersburg, Russia. 2017. № 5. - C. 18-25.

115. Hojamkulov S.Z., Mirzakulov H.Ch., Melikulova G.E., Usmanov I.I. Study of the process of decofluorination of extractive phosphoric acid from phosphorites of the Central Kyzylkum // Journal of Chemistry and Chemical Technology - Tashkent, 2020. № 2. - C. 37-39.

116. Khodjamkulov S.Z., Melikulova G.E., Khujamberdiev Sh.M., Mirzakulov Kh.Ch. Research of the process of decomposition of extractive phosphoric acid by sodium carbonate in the presence of sodium silicate // International Journal of Advanced Research in Science, Engineering and Technology // Vol. 7, Issue 9, September 2020 [ISSN: 2350-0328]. pp. 14912-14916.

117. Khodjamkulov S.Z., Khujamberdiev Sh.M., Melikulova G.E., Mirzakulov Kh.Ch., Shaymardanova M.A. Separation of phases formed during the process of desfluorization of extraction phosphoric acid with sodium phosphates // International Journal of Advanced Research in Science, Engineering and Technology // Vol. 7, Issue 10, October 2020 [ISSN: 2350-0328]. pp. 15192-15196.

118. GOST 20851.2-75. Mineral fertilizers. Methods of determination of phosphates. - M.: IPK Izd-wo standards. 1997. - 37 c.

119. GOST 20851.3-93. Mineral fertilizers. Methods of determination of the mass fraction of potassium. - Moscow: IPK Publishing House of Standards, 1995. - 41 c.

120. Burriel-Marty F., Ramirez-Muñoz X. Flame photometry. M., "Mir", 1972, 520 p.

121. GOST 13686-84. Table salt. Test methods. - Moscow: IPK Publishing House of Standards, 1984. - C. 8-10.

122. GOST 24596.4-81. Feed phosphates. Methods for determination of calcium. - M.: IPK Izd-wo standards. 2004. - 3 c.

123. GOST 24024.12-81. Phosphorus and inorganic phosphorus compounds. Methods for determination of sulphates. - Moscow: Izd-wo standards. 1981. - 4 c.

124. GOST 22275-90. Apatite concentrate. Technical conditions. - M.: Izd-wo standards. 1991. - 18 c.

125. GOST 24596.7-81. Feed phosphates. Methods for determination of fluorine. - M.: IPK Izd-wo standards. 2004. - 5 c.

126. GOST 20851.4-75 Mineral fertilizers. Methods of water determination. - M.: IPK Izd-wo standards. 2000. - 5 c.

127. GOST 18995.1-73. Chemical liquid products. Methods of density determination. - M.: IPK Izd-wo standards. 2004. - 4 c.

128. GOST 10028-81. Capillary glass viscometers. - M.: IPK Izd. of standards. 2005. - 13 c.

129. GOST 24596.5-81. Feed phosphates. Method for determination of pH of solution or suspension. - M.: IPK Izd-wo standards. 2004. - 2 c.

130. Handbook of X-ray spectrometry. / Eds. Van Grieken R.E., Markowicz A.A. - New York: Marcel Dekker Inc. 1993. - 984 p.

131. Zschornack G. Handbook of X-ray data. - Berlin, Heidelberg: Springer- Verlag. 2007. - 969 p.

132. Downs R.T., Hall-Wallace M. The American Mineralogist crystal structure database // American Mineralogist. 2003. - Vol. 88. - pp. 247-250.

Printed by Books on Demand GmbH, Norderstedt / Germany